Into the
Night

INTO THE NIGHT

Tales of Nocturnal
Wildlife Expeditions

Edited by
Rick A. Adams

University Press of Colorado
Boulder

© 2013 by University Press of Colorado

Published by University Press of Colorado
5589 Arapahoe Avenue, Suite 206C
Boulder, Colorado 80303

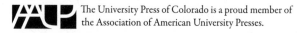 The University Press of Colorado is a proud member of
the Association of American University Presses.

The University Press of Colorado is a cooperative publishing enterprise supported, in part,
by Adams State University, Colorado State University, Fort Lewis College, Metropolitan
State University of Denver, Regis University, University of Colorado, University of Northern
Colorado, Utah State University, and Western State Colorado University.

∞ This paper meets the requirements of the ANSI/NISO Z39.48–1992 (Permanence of Paper).

Library of Congress Cataloging-in-Publication Data

Into the night : tales of nocturnal wildlife expeditions / edited by Rick A. Adams.
    pages cm
    Includes bibliographical references and index.
    ISBN 978-1-60732-269-6 (hardback) — ISBN 978-1-60732-270-2 (ebook)
    1. Nocturnal animals. 2. Wildlife watching. 3. Biology—Fieldwork. 4. Natural history—
Fieldwork. 5. Biologists—Biography. 6. Naturalists—Biography. 7. Scientific expeditions. I.
Adams, Rick A. (Rick Alan)
    QL755.5.I57 2013
    591.5'18—dc23
                            2013022974

Design by Daniel Pratt

22   21   20   19   18   17   16   15   14   13   10   9   8   7   6   5   4   3   2   1

To Jasper,

my canine companion of
eighteen years, who was at my side
through most of my adventures—RA

Contents

Preface

Loren Eiseley, the great naturalist and insomniac, wrote: "[B]ut in the city or the country small things important to our lives have no reporter except as he who does not sleep may observe them. And that man must be disencumbered of reality. He must have no commitment to the dark as do murderers and thieves. Only he must see, though what he sees may come from the night side of the planet that no man knows well. For even in the early dawn, while men lie unstirring in their sleep or stumble sleepy-eyed to work, some single episode may turn the world for a moment into the place of marvel that it is, but that we grow too day-worn to accept."

Indeed, nature's nightly marvels linger unfamiliar to most people. However, for those who choose to enter it, the night world reveals unexpected delights. Diminished light sharpens our nonvisual senses. Our attention to sounds and smells becomes piqued, offering intimate encounters with organisms that sweep through the night as easily as we navigate by day. Undeniably, immersion into the night world significantly broadens our perspective, even for those explorers who are seasoned biologists and naturalists. This book is a compilation of narratives from professional field scientists and naturalists who have found a driven magnetism within the nocturnal world. These prominent authors weave together accounts of the experiences they had working days and nights on very little sleep as they trekked through wild areas across the globe. Readers witness moments of discovery and astonishment, the compelling urges that push investigators through the dangers and challenges of conducting field studies in remote and unforgiving habitats.

These intimate essays encompass the surrealism of a sea ablaze with bioluminescent algae, avoiding the wrath of an African bull elephant, the experience of being bitten below the belt by a large and highly venomous red ctenid spider, unexpected confrontations with North American bears, cougars, and rattlesnakes, unconditional acceptance by a family of owls, dodging erupting volcanoes and hurricanes on Caribbean islands, shaking through nighttime quakes on the Pacific rim, and swimming through stratified layers of feeding-frenzied hammerhead sharks in the seas of the Galapagos Islands.

It is my hope that readers will gain insight into the world of field research being conducted by genuine biologists rather than the skewed portrayals sanitized and packaged for the audiences of Discovery Channel, Animal Planet, and Hollywood movies. For every author in this compilation, there are hundreds more in the field working in uncomfortable and dangerous conditions because they are driven by an intrinsic and profound passion for scientific inquiry and understanding. It is our intent to honor their pursuits with this insightful book and to reveal the rarely observed world of nocturnal field research.

Rick A. Adams

Into the
Night

ONE

Waiting for Long-eared Owls

STEPHEN R. JONES

I spent my first night at Pine Lake, a peaceful oasis in the Nebraska Sandhills, twenty years ago. I pitched my blue dome tent in a hillside grove of ponderosa pines, where I could gaze out across the water to the empty grass-covered dunes that rolled and tumbled toward the eastern horizon.

At first glance the mile-wide lake seemed somewhat forlorn, with its murky, leach-infested water, surrounded by rickety red picnic tables scattered across patches of mowed pasture grass and aromatic outhouses buzzing with oversized flies. The hills west of the lake sprouted plantation rows of midsize pines and red cedars—most likely a Civilian Conservation Corps inspiration from the 1930s. Between the dirt entrance road and the eastern shore, Nebraska Game and Parks had even installed a metal swing set and a little merry-go-round.

But the lake met my first requirement for prairie camping, solitude. Clearly humans had been here, and recently, but on this late-May evening none were around. Within minutes of setting up camp, I noticed

DOI: 10.5876/9781607322702:c01

the cottony sensation and faint ringing in my ears that signal escape from the perpetual background noise, the subliminal urban drone, of modern life. As the pine shadows reached out toward the water and the cottonwoods along the shore shimmied in the evening breeze, I felt the euphoria that comes from being alone in a semi-wild place.

I sat in the fragrant pine duff watching rafts of ducks and white pelicans glide across the lake and listening to the metallic chattering of marsh wrens in the cattails below me. Flashy yellow warblers and orchard orioles flitted through the willows along the near shore, while a handsome redheaded woodpecker hammered away on the silvery trunk of a dead cottonwood. At intervals, a pair of long-billed curlews wailed out warnings in the grassy uplands behind me. I heard a vague snort, like someone sneezing, and looked around just in time to see a graceful doe hoist her snow-white tail and bound away into the woods.

I brewed a mug of coffee and then alternated sips and nibbles of a piece of dark chocolate as the sun sank behind the pines and melted into the dunes. A family of coyotes off to the south heralded the moment with a rousing chorus of yips, squeals, and howls. A second family chimed in from across the water. As the first stars burned into the indigo sky, two great horned owls landed in a ponderosa above my tent and hooted me to sleep.

It was the owls, I think, that turned the trick. I had been looking for a home base in the Sandhills, a quiet retreat where I could camp out, track breeding bird populations, and immerse myself in prairie life. I study owls, and I had learned long ago that owl omens are worth heeding.

Almost every culture, during some period of its development, has revered owls as bearers of wisdom or feared them as messengers from the other side. Traditional Ojibwa stories describe how the souls of the dead must pass over an "owl bridge" to reach the spirit world. The Northwest Coast Indians say that a hooting owl portends death. The Cheyenne word *mistae* means both "spirit" and "owl." The scientific name for the burrowing owl, *Athene cunicularia*, derives from the

1.1. Fiery sunset at Pine Lake

Greek goddess of wisdom, Athena, who is often pictured with an owl perched on her shoulder. Lakota warriors carried burrowing owls into battle, believing the owls' strong medicine would repel enemy arrows.

Today, we tend to characterize such beliefs as quaint superstitions. However, anyone who has worked with owls will tell you that their aura of omniscience is well earned. Materializing and vanishing at will, owls appear wise in the way they calmly watch us. As top-rung predators endowed with supersensitive sight and hearing, they quietly take command of their surroundings, seeming self-composed and aloof.

And, for whatever reason, we sometimes become aware of them during times of grief. Many of us have heard stories of owls visiting a friend or relative after the death of a loved one. I've had this experience. I was sleeping in my mother's house in Palo Alto two nights after her death when I heard loud hoots and wails outside my bedroom window. Astonished, I recognized the hoots as those of a northern spotted owl, a threatened species that seemed entirely out of place in a suburban

1.2. Long-eared owl

backyard. A female owl and her fledgling must have flown onto the patio sometime during the night and were perched in a wisteria bush ten feet from where I slept. When I recounted this incident to a friend, she said the same thing had happened to her after her mother's death. She had gone walking alone in the Ohio woods, and a barred owl had flown over and perched on a branch right beside her.

Since I first became seriously aware of owls more than thirty years ago, they have come to me time and again, especially when I'm alone. I remember the tiny flammulated owl who hooted beside my tent in Colorado's Indian Peaks Wilderness; the northern saw-whet who caressed my hair with his talons in Boulder Mountain Park; and the great horned owl who joined me one frigid night along Nebraska's North Platte River, perching placidly on a bare cottonwood limb as Comet Hale-Bopp flared across the sky.

Each of these encounters left me more alert, more receptive to nature's gifts, and happier to be alive. In a way owls have provided a portal to a deeper connection with nature. The wisdom they have passed on is difficult to characterize, but it runs deep.

So when I heard those owls hooting above my tent at Pine Lake and found them there again at dawn, silently watching me, I decided to stick around. I kept coming back, and over twenty years of visiting through all the seasons, I grew to know the lake and its environs better than any other place on earth.

In addition to six species of owls, I documented 103 species of breeding birds at the lake. I followed porcupines through the woods, watched a mink fish from a half-submerged log, was lulled to sleep by crickets, and awakened by loons and grebes. For two years a young wild turkey adopted me, accompanying me on evening walks and trilling me awake at dawn. On moonlit nights a curious coyote sat and howled beside my tent.

I heard the great horned owls almost every evening and saw them at dawn silhouetted against the sky. I soon learned that they knew me much better than I knew them. They seemed to have the spooky ability

1.3. Roxanne the friendly wild turkey

to distinguish me from other humans, showing little fear when I came near but fleeing when someone else walked by. While strolling among the pines, I often felt a prickly sensation on the back of my neck, and I would swivel around to see a great horned owl staring at me from a nearby tree. Looking into its round impassive eyes, I could guess what it was thinking: "You again. What are you up to now?"

I saw short-eared owls coursing over the cattail marsh at the south end of the lake. Little burrowing owls bobbed up and down on the wooden fence posts that separated the wildlife area from a neighboring ranch. Every once in a while, I'd hear the hiss-scream of a barn owl deep in the woods. On warm summer nights, the quavering wails of eastern screech owls haunted the cottonwoods at the north end of the lake, where turkey vultures huddled on shadowed branches and wood ducks clucked softly to their young.

Sometimes the serenity of this wondrous place left me weak-kneed and trembling with emotion. I would stand in the dunes as the orange rays of the setting sun washed over the prairie, infusing the grass, trees, water, and sky with pure shimmering light. As the owls hooted solemnly from their roost in the pines, I felt I could stay forever.

Sadly, intimate familiarity with any wild place comes at a cost. Even in this protected wildlife area set amid 20,000 square miles of mostly native prairie, things were changing. After a few years I began to notice more shotgun shells littering the pine duff, more tire ruts carved into the dunes, more cottonwood logs stacked up for firewood in the picnic area.

One morning I watched a pair of European starlings evict a family of redheaded woodpeckers from its nest hole in a dead cottonwood. The starlings stayed; the woodpeckers became scarce. As the years went by, I observed fewer native short-tailed grouse and more introduced ring-necked pheasants. Interloping rock pigeons began to flutter through the picnic area.

It was the same story with the owls. I saw my last burrowing owls in 1992, just before the rodent colonies where they had nested disappeared. Short-eared owls became harder to find. They nest on the ground, and I feared that feral house cats, raccoons, and other human-adapted predators were preying on their young.

Witnessing this creeping loss of diversity left me feeling queasy and on edge. With each visit to the lake I became more possessive of its native inhabitants—the curlews and coyotes, resourceful badgers and long-tailed weasels, secretive bitterns and rails. Just seeing a rare or threatened native triggered a host of gnawing concerns. Would that same creature be here next year? Would this unique sanctuary remain protected? Or would all this wild beauty vanish before my eyes?

When I first saw the long-eared owls in April 1992, those visceral fears surged to the surface. My friend Roger and I were setting up camp in the pines when he called out to me in mock consternation, "Oh drat, I guess I'm going to have to move. I'll never get any rest with this long-eared owl staring at me."

1.4. Frosted dragonflies

She was hunkered down in an old crow's nest in the pine just above his tent. I dropped my camping gear and circled around to get a better look, almost forgetting to breathe.

I never expected to find long-eared owls at Pine Lake. These medium-sized owls have disappeared from much of the prairie region. They suffer from human disturbance of streamside thickets, where they nest, and cultivation of wet meadows, where they hunt mice and voles. The proliferation of great horned owls poses an additional threat. Wherever humans gather on the high plains, so do great horned owls. These larger, human-adapted predators compete with the long-ears and eat their young.

Long-eared owls range clear across the United States and southern Canada as well as through Europe, northern Asia, and parts of North Africa. Named for the false "ear" tufts that sprout from the top of their heads, these owls also possess a distinctly squarish, rusty facial disk. This disk helps to channel sound to their large, sensitive ears.

1.5. Long-eared owl on nest

Standing just over a foot tall but with wingspans of three feet or more, these acrobatic predators can dart through woodland thickets or course low over open meadows. They often catch their prey by "stalling out" and dropping straight down. Though quiet and reclusive, long-ears can be fierce when defending a nest. In *Life Histories of North American Birds of Prey*, naturalist Arthur Cleveland Bent wrote, "I know of no bird that is bolder or more demonstrative in the defense of its young, or one that can threaten the intruder with more grotesque performances or more weird and varied cries."

Roger and I didn't witness any of these aggressive behaviors. The female sat quietly, glaring resolutely at us as we backed away to a more respectful viewing distance. After several minutes of searching with binoculars, we found her mate lurking in another pine a few meters from the nest.

Ominously, the pair had nested within easy hooting distance of the pine thicket where the local great horned owls were brooding their young. I wondered whether the long-ears had any chance of success.

When I returned two months later, I discovered an empty nest flanked by two juvenile great horned owls.

For years I obsessed about the fate of those long-eared owls. Had they renested elsewhere, or had they been killed by their larger competitors? I listened for their barks and wails at night, searched for their slender silhouettes in the pines by day. Once in a while I'd hear a single resonant hoot and feel my pulse quicken, only to recognize the vocalist as a young great horned owl. Eventually, I abandoned hope of seeing or hearing them again.

While camping at the lake several years later, I decided to spend an entire June night wandering along the shore and through the woods. I had idled away the hot afternoon sitting in the shade reading a book of Pawnee mythology. The Pawnee trace their origins to the stars, and their creation stories express reverence and awe for that sacred period between dusk and dawn, when spirits haunt the dank air and visions come rattling out of the void.

One story touched me deeply. A hungry young man whose people had been suffering through a famine spent four days and nights sitting above a cave spring, praying and crying out for a vision. On the fourth night, he gazed at the reflection of the full moon on the water and saw the image of an old woman. He looked up and saw her sitting close by, near the entrance to the cave. She took him by the hand and instructed him in the ways of living. She said that if his people waited patiently, something miraculous would occur. The Pawnee waited for several moons, growing ever more hungry. Just as they were losing hope, the mouth of the cave opened wide and thousands of bison streamed out onto the prairie. After that time, the people lived well, and the earth was whole.

The ethereal beauty of the story awakened my longing for connection with that mystical, quiet time after sunset. I wanted to be out with the owls, to feel their presence in the darkness. Maybe I'd discover something magical, like a Cecropia silk moth, a rare yellow rail, or quicksilver moonlight on slate-black water.

I set out from camp an hour before sunset, descending through the pines to the dirt road that follows the western shore. The afternoon breeze had abated, and a pleasant coolness had settled over the water. The green hills across the way began to glow in the sunlight reflecting off a purple-black bank of departing thunderclouds, while the lake surface turned a deep electric blue. A family of crows flapped by, all cawing in chorus as they approached their roost in the pines.

Buoyed by the vibrant light, I strolled down toward the immense cattail-bullrush marsh at the south end of the lake. As the lake surface turned to glass, I watched a family of coots splashing about in the shallows and a half dozen black terns diving and skipping over the water.

Around sunset the first nighthawks appeared, making bull-like *vrooors* as they hurtled toward the ground and the air rushed through their wings. A bittern called from the cattails. His tranquil frog-like *oonk-a-lunk, oonk-a-lunk* seeped across the water and dissipated in the swirls of fine mist hugging the shore.

By the time I headed back around the west side of the lake, the owls had begun calling. The great horned owls in the pines near camp started up first, and a second pair answered from across the water. The nearby pair hooted in synchrony, first the male, *who-whoo, whoo-whooo*, then the female, *who-wh-wh-whoo, wh-whoo-whooo*, then the male again, monotonously, until night settled in.

The hooting serves two practical purposes: to warn away other great horned owls, and to cement the pair bond. But for me it always has a soothing quality, like the sound of a distant train whistle on a calm winter night. I stood there open-mouthed, reveling in their music, just barely resisting the temptation to hoot back.

During a pause in the performance, I strolled up through the woods and listened for the hissing sounds young owls make when begging for food. I found the family in a grove of pines one hundred meters back from shore. Dozens of owl pellets, oblong gray masses of regurgitated bones and fur, lay at the base of several excrement-splattered trunks. Something scrambled from one branch to another. A gut-wrenching wail and three harsh barks pierced the air, sufficient warning to have me muttering apologies while slinking back down the hill.

As I walked up toward the north end of the lake, I heard some ghostly wails in the cottonwoods—first a drawn-out, horse-like whinny, then an accelerating tremolo, like the sound a ping-pong ball makes when dropped on a hollow table. The screech owls were nesting there in an old woodpecker hole. At dusk they looked like ragged pieces of bark as they roosted tight against the trunk; after dark they became elusive shadows, and to see them would require using my flashlight, a sacrilege on this peaceful night.

I stopped in a wet meadow to watch the fireflies twinkle on and off as they floated from one dewy grass stem to another. Their flash "signature," a languid greenish-white streak following a gently curving line, suggested *Photuris pennsylvanicus*, a common yellow-gray firefly of grasslands from the Rockies to the Atlantic Coast. I scooped a male out of the air and held him in my cupped hand, admiring the delicacy of his yellow-striped wings and the intensity of the glow emanating from his white abdomen. I opened my hand and the firefly floated away in slow motion, like a feather riding the night breeze. I tracked his flashes across the marsh until they mingled with hundreds of others.

Around midnight a thunderstorm rumbled in from the west. I headed up into the woods to take shelter. I sat propped against a ponderosa as the lightning crackled overhead, the wind whooshed through the trees, and fat raindrops fell from the sky. When the shower passed and the damp air grew dead still, I shook the beaded droplets off my parka and strolled back down to the shore.

On the sandy bank beside the dirt road, pale evening primroses had unfurled their white crepe-paper blossoms, hoping to attract a night-flying sphinx moth. This "hummingbird moth" inserts its long proboscis into the flower's trumpet-shaped throat to extract sweet nectar. In doing so, the moth rubs against the flower's stamens, whose dusty yellow pollen sticks to the insect's fuzzy head. The moth flies to another flower and pollinates it, ensuring that a new generation of evening primrose blossoms will unfurl, embrace the darkness, and feed another generation of moths. The flowers stretched out wide, waiting patiently for the moth. I waited with them, settling into the sandy embankment as the crinkly white blossoms floated back and forth in the breeze. The Milky Way blazed silently overhead. A single cricket chirped drowsily in the meadow across the road. Nothing else stirred. I dozed.

When the high scream finally pierced the stillness, it startled me awake. *Chee-a-weet, chee-a-weet!* The synthesizer-like wail came from somewhere off in the pines. It reminded me of the alarm call of a great horned owl, but maybe higher pitched and a little weaker. *Chee-a-weet, chee-a-weet, chee-a-weet!* The sound grew closer, apparently coming from the edge of the woods. I heard some muffled barks and chicken-like squawks. Long-eared owl?!

I headed in that direction, but without any moonlight to illuminate the way, I kept crashing into branches and tripping over roots. I heard the scream again, about fifty meters ahead. I flailed through the trees until I found myself on the edge of the woods, gazing out at the black shadows of dune to the west. *Chee-a-weet, chee-a-weet* . . . Now the calls came from behind me, and I sat slump-shouldered in the damp pine needles until the screaming died away.

I returned to camp a little before dawn, slept for a few hours, then went looking for nesting long-eared owls. Thousands of pines and red cedars surround the lake, and any one of them could conceal an old crow or Cooper's hawk nest (owls don't build their own nests). During a full morning of searching, I found some medium-sized owl pellets scattered around the base of a couple of roost trees, but no other signs. I didn't see or hear any long-ears that day, nor again that year.

When I arrived at the lake a couple of Junes later, I found a barge loaded with heavy machinery floating out in the middle. A black plastic pipe drooped over the gunwale and serpentined through the water to the western shore. I heard a diesel engine chugging away and saw a gooey black substance spouting from the pipe and accumulating in a sandy hollow south of the pines. When I walked over for a closer look, one of the workers explained that they were dredging out the bottom of the lake to improve the fishing. I pitched my tent as far away from this industrial activity as possible and stayed just long enough to complete my breeding-bird survey before heading for greener—and more peaceful—pastures.

Down along the North Platte River, I met up with my friend Jack, a Caddo holy man. I told him about the dredging project. "Makes sense," he said. "Crazy white men trying to dig a hole in the lake."

I worried about the effects of this newest incursion on the lake's wildlife, and when I returned in October, I approached the lake with trepidation. I drove slowly along the dirt road that circles the north shore, keeping my eyes fixed straight ahead, dreading what I might find. When I pulled into the picnic area and got out of the car, I could see no heavy equipment, though the dredging had covered several acres of dunes and wetlands with tar-like muck. No one else was around, so I settled in for a few days and nights of hoped-for solitude.

On the second night, an eerily familiar call nudged me awake. When I realized what it was, I bolted upright in my sleeping bag and cupped my hands behind my ears to amplify the sound.

Hooo, hooo (on one pitch)
Hoooouuuu (descending like a sigh)
Hooo, hooo
Hooouuouu
Hooo, hooo
Hooouuouu

I couldn't believe it. I had heard the poignant courtship duet of long-eared owls on tape, but never in the wild. I closed my eyes and took it all in—the male's rhythmic, mellifluous hoots, the female's lilting replies, the rise and fall of their hollow voices in the crisp night air.

Then I understood. I knew those calls not merely from tape recordings but deep in my gut, and that knowledge surely evolved out of millions of years of waiting and listening in the darkened woods. I thought of my ancestors in Africa and Eurasia, sitting around a blazing fire trying to make sense of those otherworldly sounds. I thought of the owls, courting like this and rearing their young in humid woodlands populated by mammoths and saber-toothed cats, cooing and courting through the ice ages, the evolution of the North American prairie, the coming and going of diverse human cultures. I savored each note, tried to wrap myself around the music, inhale it, absorb it.

Over millions of years, long-eared owls have evolved a startling array of vocalizations, including barks, wails, and squawks of alarm; kazoo-like squeals around the nest; single hoots used to advertise nesting territories and attract mates; and this rare, exquisite duet. But these owls, sensitive to the presence of the always-alert and much larger great horned owls, seldom vocalize. I've spent peaceful evenings sitting within a couple hundred meters of an active long-ear nest without hearing a single hoot. So I felt doubly blessed to hear their clear voices, calling in unison deep into the night. The persistence and synchronicity of the duet suggested they were serious about nesting, probably in the dense thicket of pines and red cedars just north of where I had pitched my tent.

The following spring, while stretched out reading beneath a large ponderosa, I heard something rustling in that very thicket. A ball of rust-colored feathers poked out from behind a clump of pine needles, then an owlish head with two small ear tufts. The tufts were so short that at first I mistook the owl for a fledgling great horned, but it was far too small. It had to be a young long-ear fresh off the nest.

The owl struggled up onto a sturdy limb and rested in the sun. I gazed its way every few minutes, then returned to my book. After a while a flash of movement caught my eye, and I looked up just as the owl took flight and glided straight toward me. I held my breath as it flared its wings and landed on a bare branch just above my head.

To see a long-eared owl that close—its delicate rufous feathers ruffling in the breeze, its tawny-brown head swiveling, its round yellow eyes peering calmly down—seemed too miraculous to believe. Slightly shaken, I talked to the young owl in soothing tones, telling it how pleased I was to see it, how handsome it looked on its shaded perch. The owl hardly reacted at all, resting there serenely while swiveling its head through 360 degrees of the compass.

We shared that quiet space in the woods for more than an hour until the fledgling sailed off to the south and vanished into a thicket of scrawny pines. An outburst of squawks and squeals as it hopped around in the foliage suggested that it had company. I eventually counted three long-eared owls there, one adult and two young.

I found my friend Chris sitting at a picnic table down by the lakeshore. I told him about the owls and asked if he'd like to see them. We sneaked back to the big tree and sat down together. The owls went ballistic. With frightened squawks and barks, they exploded out of the pine thicket and flapped wildly away.

I returned to the lake the following October, pitching my tent in the usual spot. The next morning I rose an hour before sunrise, walked over to the big ponderosa, and sat there meditating as the pink glow of false dawn tinted the eastern sky. Just as the shapes of the trees had begun to resolve in the gray light, an owl glided by, nearly brushing my face with its silent wings.

That evening I stood alone on the west shore as the sinking sun set the golden cottonwoods and russet-red prairie on fire. Groups of white pelicans and cormorants creased the placid water, while metallic green damselflies floated from one shaggy cattail stalk to another. I heard a ripple in the blue, a rolling, pulsating call, and located a flock of several hundred sandhill cranes circling high overhead. The trum-

1.6. Cormorants in the moonlight

peting intensified as the cranes scrambled into two ragged waves and sailed south.

After sunset the coyote families exchanged yips, and the owls began vocalizing from all directions. I counted nine calling individuals—four great horned owls, two eastern screech owls, and three squawking long-ears off behind my tent. I wandered up that way and lay down in the pine needles, gazing up at the sky.

An owl barked off to my left. I hooted twice, very softly. The owl responded with shy barks and wails. I hooted again and the owl wailed back. I wasn't sure what to make of this exchange, but every time I hooted, the owl responded, and we conversed in that manner until the cold night air settled in and the last daylight drained away from the woods.

Suddenly two of the owls hung right overhead, fluttering like giant bats. I could just discern the silhouettes of their round heads twisting around to peer down at me as their wings flailed away, struggling against gravity. I felt the heat of their eyes, probing and questioning.

1.7. Curlew at dawn

They hovered there for a second or two, nearly within reach, then vanished. All the stars came out and the ponderosas began to shiver and sway. I snuggled up in my sleeping bag and drifted off to the creak and groan of the trees, the rush of the wind, and the distant calls of hunting long-eared owls.

When I awoke at dawn a dense fog had enveloped the lake and woods, softening the contours of the dunes, amplifying the cries of the wild geese and curlews, coating the bending grasses with droplets of glistening dew. In this fresh-made eiderdown world, each breath felt like a caress, each footstep, a precious gift.

TWO

African Nights among Fruit Bats, Fig Trees, and Elephants

Frank J. Bonaccorso

Imagine you are in a car in the blackness of night in wildest South Africa. The engine is off, and you and your companions sit quietly. You peer out of the window at the dead-end dirt road ahead. You hear grunts, howls, roars, snorts, and other animal noises from the direction of nearby Shingwedzi River. You hear branches snapping. Suddenly an ear-piercing, trumpet-like shriek raises goose bumps all over your body and your hair stands on end. You are aware this region is known for large and dangerous animals, including lions, leopards, Nile crocodiles, African buffalo, white rhino, and spitting cobras.

You shine your flashlight toward the dense vegetation and recall the scene in *Jurassic Park* when the T. rex emerged. But tonight what emerges is a faintly illuminated representative of the largest land animal on earth today—a bull African elephant with sizeable tusks. It appears to be the width of the aluminum can you are sitting in. The bull strides under the mist net you have suspended twenty feet aboveground in the lower canopy of a lone fig tree—and heads

DOI: 10.5876/9781607322702:c02

2.1. Author beside a sycamore fig tree used for generations as an elephant rub

straight toward you. The only object between you and certain death is the trunk of a fig tree.

The elephant seems about to continue around the fig tree toward you. Suddenly it stops and begins to rub against the tree trunk. First it rubs its neck and then the thick skin of its flank against the tree, a mere fifteen feet in front of you. It rubs up and down and sideways all at once in apparent happy contentment. You are temporarily relieved; but the elephant is so close you can hear its long protracted breaths. What would you do?

The scenario is not hypothetical. It is exactly the situation my research team and I faced in June 2004 in Kruger National Park in South Africa while we were waiting to capture fruit bats attracted to a ripe crop of figs hanging from the tree. My tale describes what we do and why every move counts when conducting nocturnal research on fruit bats in Africa. I share the unnerving and potentially dangerous

2.2. African elephant on the road

situations I have encountered while conducting research in a land full of carnivores always on the lookout for an easy meal. Indeed, it is the combination of practical experience, instinct, common sense, and luck that has allowed me to wiggle out of life-threatening predicaments— and that intrigues and draws me to Africa year after year. Some people jump out of airplanes, others attempt to climb Mt. Everest, some drive Indy racecars or play rugby; I seek wildlife encounters for my adrenaline rush. I am drawn to these places by both the science and the mystique of the unknown while working in a truly wild place—and the risks add to my enjoyment.

Let's continue my tale of the elephant at the "rubbing post." Our group sits still as can be in a right-hand-drive Toyota SUV while an unpredictable and highly dangerous twelve-foot-tall, five-ton behemoth soothes an itch right in front of us. What it will do next I haven't a clue. I derive some comfort from the trust I have in my companions.

On the right, in the driver's seat and even closer to the elephant than I am, is Professor John Winkelmann, my longtime field colleague from Gettysburg College. Behind us in the next row of seats are Emily, a student from Gettysburg, and Vilson, a game ranger and by far the most experienced person among us in dealing with elephants. In the third row of seats is Gavin, another student from Gettysburg.

I confess that whenever we encounter elephants, John usually is much calmer than I. However, we usually meet elephants in daylight, spot them at a safe distance, and have the car engine running, ready to bolt if things get nasty. On this occasion, I notice John growing more and more nervous every second. John whispers to Vilson, "What do we do?" Vilson, in his broken English, says, "Shine big light." The "big light" is a 3-million-candle-power spot lamp that is buried in our gear in the area behind Gavin. Gavin immediately rummages through the equipment—the ice chest, the telemetry antennae, and other assorted field gear. He finds the lamp, passes it to Emily, who passes it to me, and I give the cigarette lighter plug to John, who fumbles in the dark to get the plug in the socket. John then takes the business end of the lamp and directs it toward the feet of the pachyderm. Despite our near panic, the elephant continues to enjoy its rub. John presses the power switch and on comes . . . nothing. As panic moves in on us, I whisper, "The key! Turn the key." John complies and almost as soon as the light shines on the legs of the elephant, it stops rubbing. We hold our breath as the enormous beast pauses, and we cringe in terror as it takes a step toward us, followed by a second step—then suddenly the elephant veers to our right and with calm dignity walks past the driver's side of the car and continues along the road away from us. We let out a collective sigh and breathe easier.

Vilson knew from experience that an elephant will usually leave an area if big bright lights are shone on a part of its body—but not if the light is shone in its eyes, which will blind it into panic. If Vilson had not been there, we probably would have turned on the engine and tried to back out of the cul-de-sac, likely enraging the elephant into chasing us. We are now believers in our $35 spotlight and resolve to have it plugged in and always ready at night from now to eternity. We chatter about how big the elephant appeared and how small and powerless we had felt even as we were awed by the immensity of the animal and thrilled to see it rubbing against our fig tree.

Within a couple of minutes of the pachyderm's departure, John turns the spotlight on the mist net. Bingo! A fruit bat is caught in the net, struggling to get loose. John asks Vilson to make sure the elephant is really gone. Releasing the safety on his automatic weapon, Vilson steps from the car and walks up the road a few steps, carefully listening for sounds that might reveal the movement of an unseen animal as he shines his light up the road. "Elephant gone," he says. Upon his word, the rest of us are out of the car. Already, John and I have our leather gloves on; we move beneath the center of the suspended net. Emily and Gavin untie the ropes that secure either end of the elevated mist net and slowly lower it. Still a little nervous, I ask Vilson if he is sure the elephant is gone. "Elephant gone," he replies with a toothy grin and a giggle.

John catches the bottom of the net coming down. The bat is not too badly tangled, and it remains calm while John and I extricate it. Our first objective is to remove the net from the bat's claw-like toenails. Once one foot is free, John holds that foot to prevent the bat from grabbing the net again. We disengage the second foot. Then I work the net over the back toward the head. With the net over the head, we free the wings, one at a time. After the net is pulled over the head and past the jaws, the bat is free, and I am holding 100 grams of confused fruit bat. We place the bat gently into a soft cloth sack, and I hold it as John and I move back to the car. Emily and Gavin hoist the net back into position just below the canopy of the fig tree. Vilson once more scans the shrubs and undergrowth with the spotlight until everyone is back in the car. Things are going very well, and it is not yet 7:00. Inside the car, we shut all the windows before I open the bat bag in case it squirms free.

Our first goal is to determine which of two similar species, Peters's or Wahlberg's epauletted fruit bats, we are now holding. Because these species are similar in fur color and overlap in body size, the best way to identify a live specimen is to pry open its mouth and see if it has one or two palatal ridges on the roof of the mouth behind the last molar tooth. If there is one post-molar ridge, the bat is a Wahlberg's; but if there are two, it is a Peters's bat. Clearly, there is a single palatal ridge behind the last molars, confirming our first bat is a Walberg's epauletted fruit bat. (Epauletted bats have pouches in their shoulders that contain large showy patches of white fur, which they flash to attract mates; the glands of the males are well developed to provide perfumed scent as a signal to females.) We then examine the genitals and teats to determine the gender; this individual is female. When we shine a dim light from behind a wing, it is evident that the joints in the fingers (epiphyseal joints) are fully ossified. In simpler terms, bone has replaced the cartilage that would be found in a juvenile bat. If cartilage remains, light passing through the joint appears translucent, but with bone no light shines through. We have an adult. We keep the bat in the cloth bag for now.

It is time to scan the net again. John powers up the spotlight, but before he can shine it on the net, the light goes dead. Apparently we have blown a fuse in our rental vehicle. We will search for a replacement fuse the next day, but for now, John leans out the window and shines his headlamp along the net. There in the middle of the net is a bat struggling to free itself. We repeat our standard safety drill for exiting the car; however, this time Vilson has to rely on his relatively low-powered headlamp to scan for dangerous animals. He signals the all-clear with an arm motion, and the four of us exit the car and lower the net.

Our second capture is smaller in body size. Is this the second species or a juvenile Wahlberg's bat? Luckily, this bat is not badly tangled, and John and I, working in tandem, have it out of the net in short order. Back inside the car, I hold the bat with a gloved hand and, with my bare hand, carefully pry its jaws apart while John makes a visual inspection of the palatal ridges in the beam of his headlamp. After also examining the genitals and epiphyseal joints, we are pleased that we have captured a juvenile female of *Epomophorus crypturus*, Peters's epauletted bat. We all agree that capturing two bats on our first night of netting in the Shingwedzi area is sufficient. We take the net and ropes down from the fig tree, and with our two bats in their cloth bags, we head back to the research camp to attach miniature radio transmitters that will permit us to follow the movements of the bats. It takes an hour at the research camp to weigh, measure, photograph, and attach radio collars. We then drive the bats back to the point of capture for release. We expect to track these two bats through the next ten nights we will spend in the north of Kruger National Park at Shingwedzi.

A Question and a Solution: How Far Does a Fruit Bat Fly?

Kruger National Park is in trouble. The problems began in 1999 and 2000 when severe flooding affected many miles of rivers in the park, knocking down thousands of trees along riverbanks. Huge dunes of sand deposits replaced a gallery forest (a forest along a watercourse in a region otherwise devoid of trees) that had supported many kinds of wildlife with shaded cover and food in the form of fruit, leaves, and bark. We've come to South Africa to conduct research that we hope will be helpful to Kruger National Park's management plan.

The sycamore fig tree is a "keystone species" in the gallery forest in Kruger—one species that is essential in supplying resources to numerous other forms of life in its community. Each part of a sycamore fig tree supports many living organisms in a complex web of life from bacteria to insects to birds to elephants, which all depend on the tree for food, nesting material, shelter, shade, and nutrient recycling. Recently, park scientists and rangers have been worried about the complete absence of young fig trees in the park. Old trees die or are toppled by floods. If no young fig trees replace them, eventually additional floods, old age,

2.3. Fruiting sycamore fig tree

and other factors will cause a decline in the tree population, and the repercussions will trickle down to numerous animal species dependent on the resources provided by this keystone plant. Because fruit bats are important agents of seed dispersal for fig trees, we proposed to study the diet and movements of fruit bats in Kruger to learn about their role in dispersal and regeneration of the seed bank of sycamore figs.

There could be a number of reasons why young figs are not being generated in the forests of Kruger. Perhaps the pollination system is failing. Perhaps the seeds are not being dispersed effectively. Perhaps seeds are not germinating. Perhaps they are germinating but grazing by the huge numbers of herbivores in Kruger is taking out all of the fig seedlings. There are many possibilities but few answers.

Before we conducted our research, other biologists had studied the pollination system of these fig trees. Figs are pollinated by tiny fig wasps. The flowers of the fig are enclosed in the fleshy, hollow receptacle that really is a compound fruit, which is called a syconium by botanists. Hundreds of tiny flowers grow from the inner wall of the syconium. The fig wasps penetrate the walls of the syconium and, by crawling over the flowerlets, displace pollen on them. The wasps use the synconium as a place to reproduce. The scientists found that pollination was effectively being carried out by wasps.

Our bat research team would study how far fruit bats moved in a night and thus the potential range for dispersal of fig seeds. Also, by bringing some fruit bats temporarily into captivity, we could observe how long (on average) it takes a fig seed to travel through a bat's digestive system. Seeds from the bat poop would be collected, and the germination rates of both bat-passed and non-bat-passed seeds would be compared. Finally, we intended to learn if fig seedlings transplanted from greenhouses to the riverine forest would survive. Wire mesh would be placed around some seedlings to protect them from grazers; others would be planted with no protection. We would also attempt to determine which animals were destroying the fig seedlings. Was it the 120,000 impala, 30,000 zebra, 13,000 elephants, or various other grazers and browsers? My bet was that all the above animals were cumulatively consuming and trampling every last fig seedling. There seem to be too many herbivores along the river drainages of Kruger for highly edible fig seedlings to survive.

Earlier research had shown that another tree species, the baobab, had limited reproductive success because of intensive destruction by feeding animals, predominately elephants, which love to strip and eat the bark. At the turn of the twentieth century, when the nascent conservation area that became Kruger National Park was created, there were no elephants present. Within a few years of the establishment of the area, then named the Sabie Game Reserve, the first superintendent, Colonel Hamilton-Stevenson, started to find a few elephant tracks and droppings. The first elephants probably wandered into the park from Mozambique along the eastern boundary of the reserve. Nearly one hundred years later, the elephant population of Kruger exceeds 16,000, and it is growing. Ironically, Mozambique is presently being repopulated with elephants from Kruger.

Baobab trees have been heavily decimated by elephants in Kruger Park. Young baobabs grow in only a few areas, where the slope of the land is too steep for elephants to walk. Elsewhere, soon after a baobab seedling is established, the bark is stripped by elephants and the seedling dies. Nearly all of the old baobabs in the park have scars from stripped bark; however, once past a certain age the tree can tolerate stripping at the base of the trunk. The baobab population in much of Kruger is slowly aging without being replaced. Does the same fate await sycamore fig trees in Kruger National Park?

Tracking Fruit Bats: Miniature Radio Transmitters, Cotton Thread, and Nimble Fingers

The miniature radio transmitters we use on epauletted fruit bats are powered by hearing-aid batteries that will transmit for up to six weeks. We can hear a radio on a bat transmitting from about a mile away under ideal conditions (if there are no hills, large buildings, dams, dense forest, or other large obstructions between the transmitting bat and the receiver).

We attach the transmitters by strapping a collar around the bat's neck, using cotton thread cut to the neck size of each bat. The thread is passed through a hole bored into the transmitter. Plastic tubing is then fitted around the thread to create a soft cushion against the bat's neck. The thread does not constrict the bat's throat or obstruct breathing and swallowing. We use cotton thread because over a period of weeks it will rot, so the radio will fall off the bat.

It takes nimble fingers to tie a knot in the thread and avoid the bat's sharp teeth. A bat handler wearing soft leather gloves tries to restrain the bat to prevent it from biting the person tying the knot. Fortunately, epauletted fruit bats are very docile and usually remain calm while being handled.

The final step in attaching the transmitter is to glue the bottom surface of the transmitter to the fur of the bat. This prevents the radio from slipping to the underside of the neck. It's essential for the best transmission that the radio and its trailing antennae are maintained on the back of the animal. Our radios have a feature called "position sensitivity," which means that when the fruit bats hang head down in the normal resting posture of a bat, gravity pushes a mercury bead over a switch inside the radio and causes it to pulse waves at a slow rate, about once every two seconds. When a bat flies, its body is horizontal, and the mercury rolls over the switch, causing the radio to double its pulse rate. Thus, from the pulse rate, we can easily tell if the bat is flying or roosting.

To track and locate a bat by radio, we use a receiver and a directional antenna. The receiver hangs around the neck of an observer. Standing still, the observer holds the antenna (attached by a cable to the receiver) and sweeps it along a horizontal plane by pivoting the arm at the shoulder joint. The transmitted signal from the bat will be strongest when the antenna points directly at the bat. A dial on the receiver indicates the strength of the signal, which we use to estimate the distance to the bat. Once this direction is determined, the observer then uses a compass to take a bearing on the bat. In our protocol, this process is repeated once each time the bat goes into a roost or once each minute when the bat is flying. Epauletted fruit bats usually make brief flights of one to four minutes when searching for a fig, but they make longer flights when commuting greater distances between distant feeding trees or between the day-roost and feeding trees.

Five Bats, One Herd of Elephants, One Lioness, and a Billion Glittering Stars

John and I returned to Shingwedzi in July 2005 for our second field season of research. We were amazed to discover that almost no fig trees along the Shingwedzi River were bearing fruit. In June of the previous year, many trees were covered with ripe figs; however, that year was wetter than average, whereas 2005 was extremely dry by comparison. Also, that year we had arrived at Kruger a month later into the dry season.

We drove past several concentrations of fig trees in search of one bearing ripe fruit and suitable for rigging a mist net, but found no fruiting trees on our first day. The following day, I suggested that we check an isolated giant of a fig tree at the Babalala Picnic Ground nineteen miles north of Shingwedzi Camp.

John and I drove to Babalala with our old friend ranger Vilson Tenda. At Babalala, we saw a single immense fig tree with a ten-foot-diameter girth in the center of the tourist picnic area. An open-air structure with a thatch roof encircles the trunk of the fig tree. The thatch protects customers from debris dropped by birds feeding in the tree and provides shade. (Midday temperatures even in the wintertime on a sunny day can surpass 90°F.) A wooden rail "fence," eighteen inches high, encloses the area. Outside the fence and some distance away are a pump and a water tank frequently visited by elephants. In the low-lying ground near the water tank is a spring that creates a small marsh; grasses and trees are green even in the midst of the dry season, and the area is a frequent feeding and watering site for elephants, buffalo, and other grazing animals and their predators.

We arrived in late morning. Scores of fruit-eating birds were flying in and out of the fig tree, which to our delight was bearing an immense crop of tens of thousands of figs. Some of the figs were peach colored, indicating advanced ripeness. From beside the tree, we watched brown-headed parrots, green pigeons, crested barbets, black-eared glossy and Burchell's starlings, and gray hornbills. Larger birds gulped down figs whole, and the species too small to swallow an entire fig pecked at them bit by bit. We were in luck and had found our netting site! We let the man and woman who operated the food concession know that we would come back just before sunset to rig our mist net. Then it was time for the celebration: chocolate ice cream bars from the deep freeze for each of us.

2.4. Babalala rest stop, Kruger National Park

We returned that afternoon with two cars. There were six of us in our research team: me, John, three student volunteers—Brian, Jay, and Chris—and Jay's mother, Liz, who was helping us with fieldwork. Just beyond the fence, several breeding female elephants and their offspring were grazing or drinking at the water tank. Ranger Vilson was visiting his family for the evening, so we assigned Liz to keep watch on the elephants. Then we unpacked our mist net, light nylon rope, and an irregularly shaped rock. To set up the nets, we first threw two ropes (eighteen feet apart, which is the length of the mist net) over a branch of the fig tree. The nylon line was tied around the rock, and Chris, Jay, and Brian took turns throwing the rock forty feet up over the fig branch bearing numerous clusters of ripe figs. After only a few tries, we had our two ropes in good positions. The ends of the mist net were then tied to the ropes and stretched apart with enough tension to hold the net open. While we were setting up the net, Liz called out to us that some of the elephants were wandering close to the picnic area, browsing on leafy trees and grasses in the marsh. In addition, a large herd of African buffalo was now in the area and grazing the marsh. Several times Liz cautioned that the elephants were moving closer, but they remained calm and seemed interested only in feeding. An adult elephant needs to feed about eighteen hours each day to take in enough plant material to keep an energy balance. We were comfortable with the elephants as dusk approached, but we were all concerned about how close they or the buffalo might come during the night. Also, we had to keep watch for leopards and lions. I cautioned everyone that because Vilson was not with us, we needed to remain alert, listening and watching for animals beyond the rail fence.

As night approached, we wondered if fruit bats would come to this isolated fig tree. The nearest watercourse with other sycamore figs was nearly two miles away. Although numerous birds fed at the fig, many of them were habitual visitors to picnic areas, some even developing habits of stealing food or begging from people. But would the epauletted fruit bats know there was a fig tree with a huge crop of ripe figs at this isolated spot? No other fruiting trees of any type that might attract a fruit bat were evident to us in the surrounding savanna and marsh. Could fruit bats detect an odor plume of ripe fig fruits from two miles away?

Having learned the utility of a powerful spot lamp last year, we came equipped with two battery-powered spot lamps for this field season. Furthermore, each lamp could operate on rechargeable batteries rather than on a vehicle cigarette lighter. We rotated the watch duty early in the evening. The elephants and buffalo continued to graze and browse south of the picnic area. I felt comfortable with these animals as long as they maintained a distance of more than fifty yards from the fence. As daylight faded, we listened to the sounds of these relicts of the Pleistocene epoch moving through the grasses, snapping branches, and in the case of the elephants, even knocking over sizeable trees to get to desirable bark and leafy material.

In July in this region, sunset comes at about 5:30. I never tire of watching that flaming ball of gases reach the horizon and disappear. The western sky was cloudy that evening, incandescent with pink, purple, blue, and orange. By 6:00, it was nearly pitch black, and in the beams of our headlamps we could see fruit bats flying around the fig tree. At first only a couple of bats were flying among the branches, but as time passed, more circled the tree. Occasionally one landed to pick a fig and carry it by mouth to a nearby perch at the top of the fig or to another tree ringing the picnic area.

We'd been waiting and watching in the dark for forty long minutes, under the constant stress of keeping watch on the big animals outside the fence—and still we had no captures. We continuously patrolled half the perimeter of the picnic area with a spot lamp. Periodically we switched on the lamps and scanned beyond the fence. Chris called me over, somewhat nervously, and asked what could make a high-pitched repetitive four-note shriek. He shone the spotlight on a pair of bright eyes, large orange orbs, about three feet above the ground. I took the light and walked with Chris along the railing to give him a closer look. From that vantage point I could show him that the eyeshine was from a small bird, a fiery-necked nightjar, perched on a fence post.

Occasionally an elephant or a few buffalo wandered close to the fence, and we shone a spotlight on the animals until they reluctantly moved away from the light. In future nights at Babalala, I grew to appreciate the elephants' presence, even preferring to have them around because I felt that calm elephants meant there were no lions hunting in the immediate area. Elephants are noisy and obvious, but lions are stealthy, secretive, and above all quiet!

At 6:40, Jay scanned the net with his headlamp and shouted, "We have a bat!" All hands except those on the spot lamps ran to the net. The bat was soon in a holding bag. John organized a crew to "process" it. Chris and I took spotlight duty. Each of us had a specific task. Jay was the "bat handler" because he excelled in gently holding a bat while keeping it calm, and John trusted him to prevent the bat from biting him

2.5. Epauletted fruit bats, Skukuza Camp

while he fit the radio collar. Brian weighed the bats in bags, recorded data, and labeled "wing punches." This year, for genetic studies, we used a sharp punch tool, something like a miniature cookie cutter, to take two small circlets of skin tissue from the wing membrane of each bat. We wanted to learn if bats from distant river basins are genetically distinct (indicating population isolation) or similar (indicating a high degree of movement by bats between river drainages). John or I verified the species, gender, age group, and reproductive condition of all the bats. John was the master fitter of collars and radios, Liz, the photographer, recording our actions for posterity. Chris and I, who had joined the team at Kruger a week after all the others had settled into their respective roles, held the spotlights, periodically checked the net for captured bats, and were on standby to remove additional bats that might be captured. Once a collared bat was released, Chris and I would track it with the telemetry gear as long as the others continued to handle bats.

Those charged with safety spotting had to remain constantly alert, both listening for sounds that might indicate a large animal was nearby and watching for animals. But we had the opportunity to appreciate the less-threatening sounds and sights of the African night as well. Every gust of wind produced the sounds of branches rubbing against one another and the rustling of tall grasses. We heard the calls of the fiery-necked nightjar again and also those of the diminutive scops owl, which stands all of four and a half inches high. The nightjar returned to perch on a fence post, searching for insects with its oversized eyes. The distant whooping of spotted hyenas reminded us of laughter and often brought answering chuckles from our team.

It was a clear night. The stars looked magnificent, and I often took a few moments to enjoy the pleasure of being in the wilderness, far from the light pollution that blanks out our ability to see the stars in cities. The Milky Way seemed to stretch forever across the sky in a cascade of countless dots of light. In the Southern Hemisphere sky, there are many constellations unfamiliar to those of us who live in North America. However, it was comforting to see the familiar Southern Cross, which I can see at home in Hawaii, although it is not visible from the U.S. mainland. That night at Babalala, Venus and Mars were in conjunction near the rising moon.

Just as we were lulled by the peace of the night, we heard snap, snap, crash, thud—an elephant was destroying another tree and not too far off. I jumped to attention and directed the spotlight on a large female and her three-year-old offspring. We had to keep the light on these elephants for some time before the youngster would move off, and shortly afterward the mother gave up snacking to follow her nervous youngster.

By 6:50 we had two more bats in the net. Chris and I lowered the net and each began to remove one. I have handled about 15,000 bats in my life, and Chris had handled about 8. I had mine out in short order while he struggled a bit. I helped him and soon we had two bagged bats waiting in the queue for processing. Our first capture of the night received radio frequency 151.090 kilohertz (we named this bat "90") and was nearly ready for release. We usually feed collared bats honey diluted in water from a plastic pipette, a grape, or bit of fruit jam to ensure they are well hydrated and have an energy boost after their half-hour ordeal of being netted, probed, and collared by our team. Everyone enjoys feeding the bats.

After the final processing of a collared bat, we return it to a cloth bag for several minutes so that its pupils readjust to the dark, thus avoiding "night blindness" caused by our bright work lights. Finally, we return each bat to the point of its capture for release. Upon release, a bat is monitored immediately by telemetry for a short time to verify that the radio is functioning and that the bat is behaving normally. Also, we want to know in which direction the bat is moving to facilitate later searches for it. Most bats released with a radio collar fly to a nearby tree, roost for a few minutes, and try to groom off the radio collar. Soon enough they habituate to the collar—in most instances, a collared fruit bat will resume normal feeding within ten to thirty minutes because the drive to feed is very strong.

As the collaring crew worked on the second bat, Chris and I took bat number 90 to the far side of the fig tree away from the net and released it. Once realizing that its body was free of the bag, it extended its wings and took flight. We checked the radio signal from 90 and were pleased to hear a strong signal, one pulse in two seconds, indicating roosting. In fact, the signal strength and direction told us that the bat was in the top of the Babalala fig.

We netted five bats in a span of forty minutes that night at Babalala and decided that was enough. The collaring crew worked for another half hour to process two remaining bats, during which time Chris and I took down the net and ropes.

After our netting gear was stored in the car, I had an unexplained intuition to drive the roads to the north and east of Babalala to see where the elephants and buffalo were heading and if any other threatening animals were near. I informed the collaring crew over my walkie-talkie that Chris and I intended to scout the surrounding roads while they finished work. I slowly drove out the entry road to the S-56 dirt road that runs east-west of the picnic area. Chris operated the spot lamp, scanning the vegetation on each side of the road. Turning the car to the right, we proceeded to an asphalt-sealed road. As the car rounded the corner, a distant hoary figure loomed just beyond the effective range of our headlights. Driving closer, we saw a single lioness walking toward us on the highway. I turned off the engine but left the headlights on. As the lioness approached within one hundred feet, she paused, then, appearing somewhat irritated, walked off the road and headed through the thorn thicket toward Babalala. I picked up the walkie-talkie. "We have a single lioness moving off the highway and toward you. Get in your car now." Brian replied, "We copy—moving to the car."

In the thirty seconds it took Chris to turn the car and return to Babalala, the others in our crew had scooped up the last bat and all the collaring materials and were closing car doors behind them, electing to finish work on the last bat inside the car. That concluded an exciting night.

What We Have Learned

During our two years in the field, we have learned a great deal about fruit bats: their movement patterns, feeding rhythms, roosting behavior, and more. Of the twenty-three epauletted fruit bats we successfully radio-collared, all were females, and all but two of them were Wahlberg's bats. Why there were no males among the bats we captured is a mystery.

Typically, a female leaves the day-roost around half an hour after sunset. She usually flies directly to a fig tree with ripe fruits that often is less than 200 yards from the day-roost. However, we noted one-way movements of up to nine miles in a single night by one female. Among the tens of thousands of figs per tree in a single fruit crop, about 1,000 percent ripen each night. Thus, an individual tree offers ripe fruit for a

2.6. Female African lion

period of about ten days. Bats detect that a fig is ripe by odor, so they hover with their noses touching a fig to determine if it is ripe. If it is, the bat briefly lands on the tree, grabs the fig in its mouth, and then tugs the fig off the tree by dropping into flight. The entire process takes about four seconds. With a fig in its mouth, the bat will then fly to a feeding perch in a nearby tree or sometimes to one in the same tree. Hanging upside down in normal resting posture, the bat holds the fig with its thumb claws and takes large bites. The bolus of food is chewed, then pressed against the roof of the mouth to squeeze out the carbohy-drate-rich juices of the fig. The bat swallows the juice and ejects a "spat" of fiber and seeds, which falls to the ground.

At any given moment, large fig trees in Kruger National Park may have one hundred or more fruit bats either swarming in flight around them or hanging in the upper canopy munching on figs. The collective effort of all these bats produces a "rain" of fig spats and a big mess under the trees. The park custodians at the Skukuza tourist camp spend a lot of time each morning sweeping and hosing the areas under the trees and day-roosts in areas where tourists pass.

An individual bat spends four to six minutes in a roosting posture each time it consumes a fig. After finishing a fig, it grooms itself by licking its face and wings free of the juice and debris. The cycle is repeated as the bat flies back to the fruiting tree. A Wahlberg's bat may feed on twenty figs each hour and sustain its feeding bout for about two and a half hours. Then it is time to digest the meal and have a good rest at a temporary night-roost or back at the day-roost if it is not too far away. Less intense feeding may resume again later in the night. A bat that has fed really well on one night may delay the onset of feeding for several hours after sunset the following night.

Epauletted fruit bats, we learned, seem to be excellent agents of seed dispersal for fig trees, but the majority of the fig seeds fall in a dense pile below the parent trees or at a bat's day-roost. Most of these seeds probably succumb to predatory insects, bacteria, or mold. However, some seeds ingested by fruit-eating bats are swallowed along with the fig juice, pass through the digestive tract, and are defecated intact and viable while bats are flying. These few thousand seeds among the millions of seeds produced by a single tree in one fruit crop have the best probability to eventually germinate. In contrast, most of the seeds ingested by birds are destroyed in the feeding or digestive process. For example, brown-headed parrots crack open seeds and ingest the contents, and green pigeons have a grinding gizzard that pulverizes most of the seeds that they ingest. In the future, we hope to study successful germinations and seedling survivorship among fig trees.

All of the "big five" game animals (leopard, lion, rhino, buffalo, and elephant) are found in Kruger National Park. These are the African animals that game hunters of bygone days considered the most dangerous to stalk and hunt. Although my colleagues and I take some risk in working at night in Kruger where large, potentially dangerous animals are present, I feel much more secure working in the wilderness at night than I do walking at night in a large city or driving in heavy traffic in bad weather on an interstate highway. In fact, all of us take bigger risks while driving in a rainstorm or snowstorm than I do walking through the African veldt. My colleagues and students quickly learn to recognize behaviors that indicate whether an animal is calm or irritated. We learn to recognize potentially dangerous situations and how to minimize factors that could lead to accident, injury, or death.

One night in July 2005 when we returned to camp after a prolonged session of radio tracking bats at Babalala, we learned about the terrorist bombings in the London subways. Of course we were outraged and saddened at this news. Later Chris confided that on his first visit to Africa, he had already learned to feel quite comfortable working among lions, leopards, and snakes, and in driving among elephant herds, but that he was not going to venture from Heathrow Airport to visit London, as originally planned, on his homeward-bound transit stop.

As for me, I was soon to return to my home in Hawaii, where I would begin to count the days until I could return to Africa.

THREE

Undersea at Night in Darwin's Galapagos

CHRISTINA ALLEN

We descend into the sea as a group, shivering and holding hands and looking at each other like deer in headlights. It occurs to me, as I'm sure it occurs to each of the group members: *Why on earth are we doing this?* I think that we have finally gone too far. I wonder, *When did we cross the line separating daring from stupidity?* When the headlines come out, our effort will be seen for what it is: "Divers bait feeding sharks with their own bodies at night." I feel like a worm on a hook and can feel hungry eyes on me from the dark.

It's spring 1999, and I've joined an educational expedition to the Galapagos Islands 600 miles off the west coast of Ecuador. A few days earlier, on the main island of San Cristobal in the archipelago, I stumbled off the plane with my group, dazed and sweaty after a long trip, into a tiny, disorganized, and nearly empty airport. Our team, the GalapagosQuest team, is a motley crew, ranging in age and experience from a hip young computer geek to a Ph.D. anthropologist, a classroom teacher, a wry gray-haired veteran of the film industry,

DOI: 10.5876/9781607322702.c03

3.1. Galapagos Islands

and me, a biologist. Hard to imagine we will soon be the best of friends.

Our mission is to get our heads around the complex environmental issues surrounding the Galapagos and make a statement about how the islands have fared since Darwin's visit in 1835. But we are far from alone in this endeavor. About 1 million schoolkids all over the world are directing our every move. By voting on our website from their classrooms, they decide where we go and what we do each day. As we explore the exotic and unique ecosystems that inspired Darwin, we are like a traveling news team.

Our base is the *Samba*, a small boat that will take us around the sea and to the islands on a rigorous schedule of interviewing locals, collecting visuals, and writing. Every night, usually around 2:00 a.m., we upload our pieces to the Internet via satellite. Each morning, we rise puffy-eyed and confused like nocturnal creatures forced to come out into the glaring sunlight. Our day has barely begun, but students have already bombarded us with hundreds of questions.

Lights in the Darkness

On our first night out, I finish my reports at about 10:00 p.m. and walk out onto the front deck of the *Samba* as it rocks gently back and forth, tethered on its anchor. Heavy fog has settled over the water, and it's so dark that all divisions between water, air, and land blur into an inky black soup. Bleary-eyed and stiff from writing, I rub my eyes and peer into dark space. Without my dominant sense of sight, my other senses come alive. I feel my way along the wet metal railing, tasting and smelling the salty breeze as the pulse of the water lapping against the boat fills my head with a hypnotic rhythm. Even the self-defining envelope of my skin seems to disappear in the warm wet air. I feel expansive, surreal, and at one with the night.

As my eyes adjust to the dark, I notice tiny blue-green flickers of light at the edge of my vision. At first I think I've been staring at the computer screen too long, but the longer I look, the more the darkness below sparks to life. Every time I try to focus, to chase down a particular spot, it disappears. Soon my eyes tire from the strain and lose focus. It is only then that the whole scene reveals itself. Suddenly I see shape and movement and realize that I am standing above a gigantic wriggling mass of glowing, darting living things. As a scientist, I have been trained to be skeptical, but as I watch the water, my mind relaxes and accepts what I am seeing.

I see that the seething mass below me is an enormous school of fish, milling about under and around the boat. The school is lit up with such clarity and definition that I start to see different shapes and sizes of fish as they cruise around in loose groups. I watch, mesmerized, as new players enter and exit. Like a blazing bottle-rocket, a big predatory fish whizzes into view as its smaller prey dart this way and that to escape. Next, a medium-sized ray flutters in and out of the school with measured butterfly flaps. Then a mystery animal glides in, swooping like a giant swallow with long wing-like fins. I'm looking at it in wonder, at a loss to know what it might be, when it reveals its identity by exhaling loudly through long wet whiskers. A sea lion!

For the grand finale, a large luminous shape glides in, slowly coming into focus. Its motion more than its size or shape gives it away. Slow, measured, and menacing, the shark cruises closer and closer, paying no apparent attention to the boat or all the frenetic little fish trying to get out of the way. Even from my safe perch on the bow, the sight of it inspires excitement and dread. As it passes directly below me, I can see it is much bigger than the sea lion and has an ungainly wide head that wags back and forth as it swims. In the next instant, the hammerhead swishes its tail once and is gone, leaving a trail of light in its wake.

The bright sound of my teammates' voices on the deck seems far away, tinny, and unnatural, but like a line thrown to me from another dimension, it pulls me back from the foggy depths of my reverie. Topside, my scientific thought comes flooding back, pushing aside the dreamy physical sensations of the water world. Suddenly the term *bioluminescence* flashes in my mind, like a billboard in lights, like the proud answer to a test question. But as an explanation to what I just felt and experienced, it feels weak, inadequate, a fancy know-it-all term. "Bioluminescence, single-cell algae, *dinoflagellates*" are the first words my ego wants to boast to my new teammates, so I, the "team ecologist," can impress them. But I am still enough in my bliss, my wonder at the magic of all that we don't know, that I swallow the words and just smile.

I join my teammates and we mill about on deck, coming together, drifting apart, aware of our personal space—not so different from the organisms below. We settle in the bow, each of us looking out, as the engines fire up to head north and cross the equator for the first time. The air is festive, charged with anticipation and camaraderie. I thrust my head into the wind like the figurehead of a ship with an important mission. In the next instant, glowing shapes zoom in from all directions and there are ten huge ghostly white dolphins performing a flying ballet just feet below our outstretched faces. My teammates shout in awe and wonder. The dolphins weave and dive around each other— gleaming ribbons of life. We all scream and cheer, unable to contain the joy we feel. I have an intense urge to jump in and join them, our

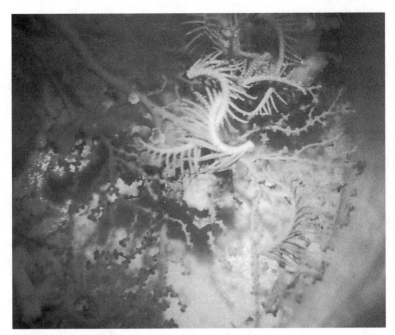

3.2. Bioluminescence

own world seeming suddenly flat and dull in comparison to their limit-less sea, our bodies awkward in the presence of their grace and freedom. I feel a bit like Alice in Wonderland as the world doubles in size to include the underwater realm. The boat that seemed so sturdy and safe now seems like a tiny chip of material bobbing on the surface of a vast and unexplored universe. Just before we cross the equator, as if nearing some invisible boundary, the dolphins peel off one by one, disappearing like vapors into the depths. The next time we look, the sea is black and empty—the luminescence gone.

An Indicator Species? Threats and Reasons for Hope

As the sun comes up the next day, I wonder for a second if last night really happened. Was I chosen to glimpse a rare and magical vision?

While I'm contemplating this mystical possibility, the real message comes to me: the realization that what I saw happens every single night, in oceans all over the world, and has happened since before humans walked the earth. Juxtaposed with this realization that what I saw is so utterly common is the shock that we still know so little about the world's oceans. The ultimate frontier in science is literally right under our noses, and while we dump our sewage into it, play—bikini-clad— at its margins, and run our motors over its surface, never really looking to see what's below, species are being lost before their secrets have even begun to be understood. Rather than a pretty scene of boats and sun playing off water, I see a huge and unknown world that holds both the secrets of the origins of life and answers to future problems.

Seeing the Galapagos Islands for the first time in daylight, I imagine Darwin also seeing them for the first time, initially from his boat, then hiking across burnt lava so sharp it cuts your shoes. From afar, the islands appear inconsequential, tiny jagged jumbles of rocks jutting out of a vast and wild ocean. Like me, did Darwin at first underestimate their importance? Even up close, the bleak and charred volcanic landscape appears lifeless, especially compared to the rich reservoir of underwater life surrounding the islands. Looking out, I realize I had expected to see Darwin's ideas obvious and in action: perfect depictions of natural selection, like an illustration from an old zoology book, with five finches in sight at once, each eating something perfectly suited to the size and shape of its beak. But I'm realizing that knowing about a scientific concept like speciation or bioluminescence is not the same as seeing it up close with its many textures and layers. Nature is, ultimately, a messy jumble that never perfectly fits the neat scientific theories made up about it. I find new respect for Darwin, imagining the vision and dogged determination it must have taken to distill his important theories from the tangle of rock and plants that are one's first impression of the Galapagos Islands.

As I am thinking about Darwin, a rather large bird flies right by me at eye level, balancing on the ocean breeze. My first thought is *seagull*, but something doesn't make sense. This seagull is almost pure black! I look in my bird book until I find it. It's a "lava gull," unique to the Galapagos. I feel better about not knowing the identity of the bird, since I would never have seen it anywhere but here—and the experience leads me on to a whole new train of questions. *How did the first seagull get here and from where? How many perished before a population took hold? How long did it take for a "regular" seagull to become a Galapagos lava gull?* I turn to my stack of field guides for some answers and again think of Darwin—with no books, no answers, only questions. Almost all animals now living in the Galapagos originally came from Peru, the nearest mainland. Most of them floated or were blown off course by storms and had the incredible good fortune to land on the tiny islands—the only land for hundreds of miles. Suddenly I see the islands not just as bleak and inhospitable rocks, but as lifesaving land, a paradise for a lucky few animals exhausted by days of battling winds and water currents.

Today, the Galapagos are paradise to more than just stranded animals. People live on and visit the islands, and what they bring and leave behind damages the fragile oceanic islands. The destruction is shocking in its speed and extent, especially juxtaposed with the creepingly slow progress of evolution, which has crafted and polished the life forms that have survived the test of the thousands and even millions of years before the arrival of humans. "Invasive exotics" now outnumber native plant species. As much as 24 percent of plants and 50 percent of vertebrate species on the Galapagos Islands are endangered. In 1859, looking at a relatively pristine Galapagos unpopulated by people, Darwin stated, "It is not the strongest of the species that survives, nor the most intelligent. It is the one that is most adaptable to change." Ironically, as visitation to the islands reaches levels Darwin couldn't have dreamed of, his words have never seemed more prophetic. Man, a relatively new species to the Galapagos, has proven to be the most adaptable and will no doubt outlive the native species perfected by millions of years of evolution. Current headlines about the Galapagos usually read like these: "Conservation on the brink," "World heritage in danger," "Longline fishing to be allowed in Galapagos," "Chaos reigns in Galapagos NP." The sights I'm seeing may soon be a memory.

Along with assessing the general state of the Galapagos Islands, our team and the students guiding us are trying to shed light on a scientific mystery. The Galapagos damsel, a.k.a. Blackspot chromis (*Azurina eupalama*), has not been seen since 1982. Guest team member and Galapagos naturalist Jack Grove, the first to document the disappearance of this native damselfish, thinks that the fish was killed off in the 1982–1983 El Niño event. During El Niño, westbound trade winds slow down and a swathe of warm water the width of the United States sloshes from its usual place near Australia and piles up on the Peruvian coast, causing a suppression of nutrient-rich cold water and a massive die-off of sea life, mainly fish and birds. The 1982–1983 El Niño, thought to be the worst in human history, caused surface ocean temperatures off Peru to rise more than 7° and rainfall to rise from a normal yearly average of six inches to more than eleven feet, causing massive flooding and long-term changes in weather patterns.

The Galapagos damselfish could be the first species in the Galapagos to become extinct as a result of a "natural event." In Darwin's day, it was widely believed that humans could not cause the extinction of marine species; the ocean just seemed too vast and human population, one-sixth of what it is today, too small. With global human population now increasing by more than 200,000 per day, times have changed and human-caused extinctions are yesterday's news. Still, the myths and misconceptions of past centuries have been replaced by new ones. Many people today undervalue the calamitous effects of natural events and, even more important, of global climate change. The possible extinction of the damselfish is especially important now because El Niño could be a predictor of the future effects of global climate change. If the GalapagosQuest team could find the missing damselfish, it would be one of the greatest underwater discoveries in decades.

The GalapagosQuest team and all the schoolchildren following along are also very concerned about another group of "fish." Though many of us fear sharks, we also realize that as top predators they are an important part of the ocean food chain. We are dismayed to learn from Jack that sharks all over the world are in danger from overfishing even in the "pristine" Galapagos Islands, which should serve as a refuge. Many scientists say that 600,000 to 700,000 sharks are killed illegally in the Galapagos each year. People catch sharks in huge numbers to cut off their fins—a delicacy in some cuisines. Often they throw the shark bodies, some still alive, back into the sea. They are destined to drown or be eaten by other sharks attracted to the trail of bloody carcasses. "Finning" was outlawed in 2004 but is said to be still rampant, thanks to a new law that allows "by-catch" to be exported. (*By-catch* is a term meaning everything caught in nets that isn't the target species. There is an incentive for big commercial fishing operations to use giant nets that catch everything from turtles and sea lions to juvenile fish. Some estimates put by-catch as high as 25 percent of total catch.) Since fishing is almost impossible to monitor closely, all shark fins illegally collected in the Galapagos can nonetheless be legally exported as by-catch.

Our team decides to go underwater to look for damselfish and sharks. I for one am thrilled about this opportunity for adventure and scientific discovery.

A Fish Out of Her Comfort Zone

From the time I was young, my parents and their friends said I swam like a fish. I carried this badge proudly my whole life, showing off with flips on the diving board, doing little dives while snorkeling, resurfacing with a showy puff of water emitted from my tube, calmly putting all my gear back on underwater in my scuba certification class. I could prove myself adept in any aquatic environment.

It's my first time diving in the Galapagos, and I am suddenly nervous. We are joined by famed underwater researchers Jean-Michel Cousteau and Richard Murphy. Though I was a "fish" when I was young, I feel inexperienced in their presence. We have come to the northern islands, Wolf and Darwin, to do some seriously advanced diving and to track down one of our main study subjects—and I am not talking about the damselfish. The currents around these tiny islands are cold and strong and carry lots of miniature sea life. Consequently, they attract fish that are followed by bigger animals like sea lions. But that's also not the reason for my nervousness. The currents attract another notorious creature . . . the seas here are literally seething with sharks.

I probably wouldn't even think about diving into a sea of sharks with anyone other than Cousteau and "Murph." They exude calm confidence and experience. They told us earlier that there is no danger from the various sharks here, that none have ever attacked a scuba diver, that they aren't feeding now, during the day, and so on. Sitting in the spacious cabin drinking a cocktail, it all sounded perfectly safe and reasonable, an exciting adventure, a lark. Now, at 8:00 a.m., sitting on the edge of a wildly bucking Zodiac, peering into the cold white-capped water, things seem very different. I'm trying to look calm and prove I can hang with the professionals, but when I flip backward into the water, I gasp, first at the cold water, then at the sight as I right myself and look down into the deep dark blue. The water is so clear you can see more than one hundred feet, and looking straight down, all I see are sharks. The first thought that comes to mind is "shark pit." I envision old Bond movies where bad guys are thrown into a writhing mass of sharks, which, like starving piranha, tear them to bits. I can't believe I'm going down there. A deafening noise confuses my senses. I start to feel claustrophobic and then get that floating feeling, departing my body. I'm sucking air out of my regulator and my mouth is dry. I realize the loud noise is my heart pounding. My eyes must be bulging because someone comes up, grabs my arm, looks right into my face, and gives me the Okay? sign.

I look around at the rest of the group, hanging in the water, waiting for me to hold up my tube to let air out and descend. I focus on my breathing: in, out, in, out. I realize we're all in this together. I look to my teammate anthropologist John Fox for help. He gives me a goofy grin, shrugs his shoulders, and starts to descend. The whole group is going down, and I'm not about to be caught on the surface by myself (images of surfers silhouetted and looking just like seals pop into my head). I don't even want to be at the back of the pack. I feel safer somehow in the middle, surrounded by my own kind. So I hold up my tube, watch the bubbles float up to the surface, and down I go, out of my element and completely defenseless. How could I ever have been so arrogant to think myself a fish?

As we descend it gets darker, as if we're swimming away from the day into night, like time has speeded up and every stroke takes us ten minutes farther from sunlight and toward the end of the day. I have the urge to turn around. My entire body is getting tense and a voice in my head is yelling, *We need the sun! We depend on the sun! We love the sun!* My skin instantly pops into goose bumps as if attempting to put on armor, a meager defense against the thick-skinned, often spiny and poisonous, residents of the deeper ocean. I look down at the neoprene wetsuit covering my pale soft body and think, *Who am I kidding?* I'm like a juicy apple, thinking I'm protected by a thin waxy film. Suddenly I'm reminded of a shark attack victim I heard years ago describing her memory of being bitten by a great white. She said she could hear the shark's teeth make a crisp *Pop!* as they punctured her wetsuit, like biting into an apple. I imagine I'd feel a lot better if I was covered with hard scales like most fish or bony spikes like a puffer or lionfish. I'm more like a big fat juicy worm hanging out down here. The only hard things on me are my fingernails and toenails. Then I imagine my whole body covered in toenails—how constricting that would feel. What if you had an itch? That's the price for protection, I guess. But then even a coating of toenails doesn't really protect fish from being eaten by bigger fish.

As we dive, light is not the only thing to disappear. With light goes color, as they are both absorbed and reflected by the water near the top.

Red is absorbed first, which is why swimming pools appear blue: the absence of the color red. Even in the middle of the day, in deep water you need giant, powerful heavy flashlights to see things in their true colors. We didn't bring lights today, so as we go down, our faces take on a deathly bluish cast. The sharks as well look grayish blue, metallic, robotic.

We continue to descend, and I start to feel claustrophobic in the dark and more than a little fatigued, my muscles still tense and my eyes open wide. Is this how it feels to live down here, constantly straining to see and afraid for your life? I know that fish and other creatures that live here are adapted for the environment, so they probably don't feel cold or tired all the time, but I do wonder if they feel afraid on some level, if they are always on edge. Just as I am wondering if fish feel fear, I am snapped back into the here and now by a big dark shape cruising into view. The shark cruises out of view again, but I am left with a sharp, tingly, super-alertness as my body is flooded with adrenaline. I look all around me to get my bearings. Below I can make out different types of sharks, organized on levels. The white-tip reef sharks are closest to me, their white-tipped fins flashing through the water. Just below them are the brownish-yellow Galapagos sharks, and even further below them are the huge and strange-looking hammerheads. All three are cruising in characteristic shark fashion, straight and purposeful, not moving a fin until the last minute, when they turn and coast again.

As we get deeper, we pass through layers of reef sharks and then Galapagos sharks with such seamless fluidity that we hardly realize they are carefully avoiding us. We see the layer below us, but it's like flying through clouds; the layer eludes us until we look up and see it solidly reformed above us, as if the last connection to the sun and the safe world above has closed. Before we know it, we are in the midst of the hammerheads, to me the most exotic and creepy denizens of the dark and deep. Their flat wide heads with eyes and nostrils on the far ends look alien, more like huge garden tools than part of a living creature. Their heads make these sharks seem even more cold and calculating

3.3. Swarm of hammerhead sharks

than other sharks. It's impossible for me to look at a hammerhead and empathize with it or try to imagine what it may be thinking. Scientists once thought that the hammer shape of the head helped the shark turn sharply without losing stability or acted as a wing to provide lift. A later theory suggested the hammer helped the shark manipulate prey. The latest theories propose that the hammer is a multipurpose sensory organ, almost like an insect's antennae. It amazes me that scientists still don't really understand some of the basic biology of such big animals. It's exciting and at the same time depressing because sharks could be gone before we know most of their secrets.

The things we do know about sharks are astounding. For instance, hammerheads, like many sharks, have sensory "pores," called the ampullae of Lorenzini, that detect electrical signals given off by living things, boats, and even ocean currents. By having these sense organs spread out over the larger area of the hammer, the sharks can sweep for prey more effectively. They can detect an electrical signal of half a billionth of a volt. The hammer-shaped head also gives them larger nasal tracts, making their chance of finding a particle in the water at least ten times greater than that of other sharks. So they are super-duper killing machines, but lucky for me they have small mouths and normally use their shark superpowers to hunt along the bottom, eating fish, rays, cephalopods, and crustaceans. They form peaceful schools of over one hundred during the day, but in the evening, like other sharks, they become solitary hunters.

As we paddle closer, the sharks slowly move away and go about their business, keeping a constant distance between them and us. It might seem strange to use the words *comfortable* and *swimming with hundreds of sharks* in the same sentence, especially after my initial panic, but their measured behavior is so precise and predictable, it gives me a sense of confidence. Before long, we're swimming into groups of sharks, testing them, astonished by the way they part to let us through and then close behind us. Perhaps it's our way of trying to communicate with a being that seems so foreign—like trying in vain to make a robot trip or cry. But they never make a mistake. It feels like we have an invisible force field around us, a bubble about ten feet in diameter that nothing can get through. We become so sure of this imaginary force field that we try to push the envelope, hiding below rocky ridges, waiting for a few sharks to get right above us before bursting up to try to get closer to them. They never let us. Ten feet, that is their comfort zone. It is also almost exactly the length of their bodies. In any case, so far there seems to be no shortage of sharks in the Galapagos.

The next day we're going to dive with the sharks again, but this time along the wall of the island rather than in very deep open water. From the start, everyone feels more comfortable, and we're laughing and joking as we splash backward off the boat into the water. The current seems especially strong on the surface today, or maybe I was just too scared and distracted to notice yesterday. For some reason, my ears won't "clear"; I feel the increased water pressure, and my ears hurt as I try to descend. I yawn and move my jaw around and finally a loud pop brings relief. When I get to the predetermined depth, I look around. I don't see the group, but there's only one clear way to go, so I swim off, sure I'll catch sight of the others around the bend, especially as fast as the current is taking me. The ocean floor is an undulating carpet of softly waving plants, but as I get closer, I see that just under the plants, it's chunky with jagged piles of pockmarked volcanic stone. Getting close to the ocean floor also shows me how fast the currents are taking me, and I grab onto a big black rock to slow myself down and take a look around.

As I turn my head sideways to look for my team members, the current rips the mask off my face. I panic as I try to open my eyes to see where I am but the saltwater burns my eyes. A vision pops in my head of my being swept out to sea on the current, grasping hopelessly at sharp volcanic rocks while I struggle blindly with my mask. Still holding onto the rock with one hand, I finally get my mask back on with the other hand and push the water out by blowing air into the mask from my nose. As soon as my mask is clear, I take a deep breath and look around. It's hard enough to see through the sides of a dive mask in ideal conditions. I feel like a horse with blinders on. On my own, and with sharks around, I'm even more nervous, opening my eyes wide and jerking my head back and forth to try to keep track of what's around me.

The exertion gets my heart pounding, and it also gives me a taste of something very important to the ecology of the Galapagos. The same strong currents that ripped off my mask are the very reason there are so many species living on the Galapagos today. There are five ocean currents that collide at the Galapagos, bringing species from land and sea to the islands from diverse parts of the mainland. Animals can float on trees or pieces of land that break free in big storms. For example, California sea lions were carried to the Galapagos from the North American coast to become the Galapagos sea lions, and the tiny penguins that were brought on currents all the way from the sub-Antarctic

3.4. Galapagos penguins

became the Galapagos penguins. Currents also bring cold water that carries nutrients and lowers the temperature of the land and sea in places, permitting a broader range of species to survive at the equator.

The currents, some colder and faster, some warmer and slower, also determine the climate of each island. And because resources like water and soil are in such demand in general, plants and animals have competed fiercely and adapted dramatically to each unique island. So you end up with a bunch of slightly different species, like the thirteen different species of "Darwin's finches" that all evolved from one type of finch blown off course hundreds of thousands of years ago. There are some 500 species of fish in the Galapagos, all evolved from mainland fish swept away by ocean currents. A stunning 40 percent of all the plant life and even more of the animal life in the Galapagos are endemic. My favorite examples are the Galapagos marine iguanas that fling their cold-blooded bodies off the sun-warmed cliffs into the icy water below, struggling against currents and heavy surf, not to

3.5. Galapagos marine iguanas

mention lack of oxygen, to graze on sparse bits of seaweed sometimes twenty feet below the surface. I imagine all the iguanas that must have arrived but starved when they couldn't find enough food on the islands. And then I imagine the first renegade iguana that was able to go so against its own survival instincts to find a completely improbable new food source in the ocean. And now there are thousands of "marine" iguanas on certain islands. Some other examples of the often bizarre endemics on the Galapagos are vampire finches, which live off the blood of other animals, and flightless cormorants with turquoise eyes that dive deep to harvest seaweed to protect their eggs from the sharp lava rocks.

As if sent to illustrate my thoughts, a ridiculous-looking puffer fish zips past, its goofy bulging eyes and tiny whizzing fins looking like cartoon features. It is also funny because one of the things the students have demanded (by unanimous vote) to see is a puffer species famous for its lovely habit of eating the poop of other fishes. I release my hold on the rocks and let myself be pulled along on the current with the puffer. There aren't many sharks right next to the island, but following the puffer leads me to some amazing bright fish darting among the rocks: dazzling rainbow parrot fish, elegant butterfly fish, and stick-like trumpet fish.

I'm always looking, but the damselfish is not here. Peeking into rock crevices at night, I would also see murky balls of slime swaying in the water. These are the sleeping nests of fish that are active in the day: slime nests. Distasteful as it may be, slime is incredibly important to fish and, surprisingly, the one characteristic that links all fish in the world. Not all fish have scales, fins, gills, or even live underwater all the time. But all fish do have glands in their skin that secrete a protein called mucin. When mixed with water, mucin makes slime. All fish have slime on their skin to protect them from scrapes and pathogens; and many fish make big puffy slime nests. I'm sure there's some clever critter down here that eats up all the slime when the fish are done sleeping in it—again, the fascinating things we could learn about ocean life are literally unlimited.

Once in a while, a shiny silver torpedo-like fish goes whizzing by, solid but aerodynamic, some sort of tuna, I guess. I'm cruising along, watching the fish, feeling more confident, independent, and capable, like I can handle anything. The island keeps curving around and the rocky wall gives me a sense of security; as long as I'm near it, I can't get lost, and I'm still sure I'll see the group around the next bend.

Time flies when you're having fun, and it seems an instant since I last checked my air gauge, which was then at half a tank. I check it again and realize I'm almost out of air. Still unconcerned, I begin to surface from this dark underworld, sure that as soon as I pop my head out of the water, I'll see the rest of the team. I surface and look around, feeling refreshed. I'm surprised to find the water choppy on top because it felt so smooth underneath. As I ride up on a swell, I get a good look around. On one side of me is Wolf, the smallest and farthest north of the Galapagos Islands. On the other side of me, only a few hundred feet away, is a tiny rocky outcrop. I not only don't see the dive team, I don't even see our ship, which is large enough to see from about a mile away.

The realization hits me like a cannonball in the chest. I'm all alone, bobbing around in water that's moving so fast I've already been swept past the small rocky outcrop. I look past the island in the direction of the current and get a glimpse of the wide-open sea, flecked with white caps. There's not another boat in sight. I feel dizzy, like I'm looking over the edge of a steep cliff. I try not to think about the huge schools of sharks, creatures of the night that are always there in the deeper water near the islands. I try not to think about how I look silhouetted from below, my dark shape bobbing around on the surface. I try not to think about what I learned last night when I did research for my article on sharks. Galapagos sharks and great hammerheads are among the few shark species that have attacked people unprovoked. *But not here,* I tell myself. *Here they are resting, not hunting.* Then, without warning, comes a sequence of physical events that I cannot control. Fast breathing, tight throat, and hot burning tears. I think, *Oh, God, this isn't happening. How did I get into this? I'm so stupid. I did this for fun and now I'm going to die.* The rocks at the edge are only one hundred yards away, but as the white water slams down on the volcanic shards, it reminds me of a cheese grater, and I know it would be suicide to try to reach them.

I've been dreading looking down into the water, fearing I wouldn't be able to keep it together if I looked down and saw layers of roving sharks. I start to argue with myself. *What would you be able to do if you did see them? Nothing, so better not to see them—it'll just make you scared.* But my imagination is even worse than reality, so I just have to look. I hold my breath and plunge my face into the water. I look down, all around. No sharks in sight. I feel better. I pull out the emergency flag I stuffed into my vest at the last minute, assemble it like a tent pole, and start waving it slowly back and forth. I don't see anyone or any ship, so I don't know how anyone would see me, but I can't think of what else to do, so I wave and wave and try to take deep breaths. Then suddenly I get an intense feeling of panic, a vision of legs, my own legs, dangling in deep water, seen from below as in the movie *Jaws*. I have to look again. I stick my face back in the water and look around. No sharks.

I'm about to pick my head up again when a dark shape comes at me fast from one side, then whips away so quickly I can't be sure what it was. My heart leaps, then starts pounding, and I'm breathing fast, almost hyperventilating. I'm looking around with my eyes bugged out, mainly to the side it came from, when it comes darting again from the opposite side, dark and fast. I see it better this time as it darts away. It's a shark, for sure, small but hard and smooth like a bullet. An image flashes to mind of a blue shark, small and solitary, from the open ocean. I pull my head up again and look around, desperate. There's no boat in sight. I'm wracking my brain for any knowledge of what to do in this worst-case scenario. I'm waving the flag frantically, breathing fast and sobbing. My whole body's shaking, and I can only imagine the signals I'm giving out underwater, but I can't help it. I look down again, and a third time the shark darts at me, again from a different direction. I envision its fin nicking my suit and starting a bleeding, feeding frenzy, all the sharks within a mile sensing the blood in the water. I'm praying to God to let me live, I'll do anything, just please let me live.

Just when I think it's inevitable that the shark is going to attack, I hear a buzzing sound and our little Zodiac comes into view. Instead of being relieved, I can't stop thinking: *Don't bite me, don't bite me, don't bite me, let me get in the boat, let me get in the boat.* The boatman smiles at me calmly as he approaches and overshoots, going past me and preparing to circle around again. "*Help!*" I yell, waving for him to come over to me. I'm sure the shark is going to slice me, bite me, and pull me under just as help is in sight. Finally, he comes up to me again and I try to scramble up onto the boat. "Hold on, let me get the ladder," he says in Spanish. I'm trying not to look like a bumbling idiot, but I'm desperate to get in the boat. "Un tiburón! Un tiburón!" I shout, but he looks at me like I'm a dumb tourist. I'm sure he's told people a million times not to be afraid of sharks. Sharks never attack in the Galapagos. I'm shivering all the way back to the ship, grateful to be alive but sheepish and humbled and definitely shaken up.

When I get on the ship, everyone is excited about the dive and his or her own experiences. I realize in an instant that this was one of the most extreme "You had to be there" moments of my life. To the others, I wasn't gone long, here I am safe, and "Wasn't that killer?" Yeah. Almost.

Eagle Rays, Fighter Pilots, and Kamikaze Boobies

The next day, I'm back in the water. Shallow water. Shark-free water. Flipping backward out of the dinghy, I hold my mask on my face and my regulator in my mouth. Compressed air in the tank allows me to breathe, my wetsuit keeps me warm, my weight-belt helps me sink, and a buoyancy compensator that fills with air helps me return to the surface. With all this technology to help me feel comfortable in the water, why do I feel so awkward and out of place?

At about ten feet deep, my ears start to hurt from the increasing pressure as we retreat below the surface. To "equalize" the pressure inside and outside of my head, I hold my nose and blow, adding air from my throat to my sinus cavities. Ahh, relief. Once my ears clear, strange sounds erupt all around me. The surprisingly loud *Crack!* of a shrimp, which stuns its tiny prey with the snap of a claw. Waves lapping on a distant shore. My own breathing through the regulator and several other sounds I couldn't begin to identify. With the bioluminescence, I felt like I had X-ray eyes, but now I feel like Superman with bionic hearing. Sound travels about four times faster in water than in air; that's how divers and other sea creatures can hear sounds produced far, far away. Some whales, for example, can communicate thousands of miles. It's exciting to think that I might even hear whales singing!

At forty feet I start feeling cold. I check my thermometer. *It's 80°? I should be warm!* I find out later you feel so much colder in water

3.6. Eagle rays

because water draws, or "conducts," heat away from your body much faster than air. Because water has more direct contact with your skin than air as it flows around you, you feel colder. I forget about the cold when I see, approaching out of the murky distance, large dark shapes. First one, then three, then five, then twenty or more. They appear in formation like a squadron of flying carpets. As they near, I can make out their white underbellies and dark speckled tops. A huge group of spotted eagle rays! I duck behind a rocky outcrop and hold my breath, waiting for them to swim right over me. When they are directly overhead, I kick my fins hard, and the next thing I know, I am right in the middle of the group. They are above me, below me, and to each side. As I continue to kick hard, I examine the gill slits along the creamy white bellies above me, then the indistinct borders around the spots on the dark backs below me. Together, we glide through the water, flying. I'm breathing heavily and feeling rather Zen, losing myself in the moment. Just as I feel I am becoming a spotted eagle ray, the low-air alarm on my air gauge goes off, beeping madly. I have been so focused on the rays, so excited to be amid them, that without even realizing it, I've followed them down to sixty feet! I look around, in shock, not sure what to do. It doesn't even occur to me at first to be afraid for my life.

I wave like crazy to the person closest to me, making the knife-across-the-throat sign that means I'm out of air. He waves back, smiling, in his own underwater reverie. I realize with horror that I'm on my own. Now I *am* afraid for my life with every cell of my body. Will to live and sheer instinct take over. I kick straight to the surface, not stopping for anything, even a potentially life-saving "rest stop" to keep me from getting the "bends." Once I get to the surface and realize I could have died ten different ways, I kiss the deck of the boat in relief, and then, finally, feel incredibly stupid. I can't face my teammates, especially since my first "mishap" was only a couple days earlier. I decide to have a gin and tonic and pretend it never happened.

Later in the same day we're rowing a dinghy to shore to explore the land iguanas of Fernandina Island. Normal iguanas, plant-eating, sun-loving iguanas. I'm looking around at all the spiky black rocks splattered with white bird poop and out of the blue, my body flinches as bombs torpedo the surface of the water all around our boat, making a hollow popping sound as they slice through the water at eighty miles an hour. It feels like World War III. We duck down in the dinghy, cringing. The missiles are actually boobies, not bombs. We have rowed out into the water over a huge school of small fish and hungry flocks of birds (there are hundreds of birds) have congregated in what seemed like an instant. As our fight-or-flight reflex subsides, we notice pale blue-footed boobies, dark and sleek cormorants, and pelicans. The sky becomes dark with birds, and the torpedo noises continue. The birds synchronize their attacks with shrill calls; then in unison they crash into the water as shallow as three feet deep, popping up like corks about ten feet away and taking off again, all flying in the same

3.7. Sea lions at play

direction and forming a conveyer belt of blurred bodies that rises in a half circle up to about one hundred feet, where they appear to fall off a cliff in organized turns, diving and crashing and popping and flying in a methodical rhythm. We stare, amazed at this feat of engineering and cross-species cooperation.

Later I attempt a cross-species friendship as I enter the swirling fish-filled water next to the dinghy. I wait in anticipation, seeing only flashes and bubbles at first. As the froth around me clears, I come face-to-face with big friendly eyes, a whiskery muzzle, long graceful wing-like flippers and, I swear, a smile. Sea lions are not afraid or aggressive; they don't swim away, and they don't threaten. As they swoop around me, their bodies within inches of mine, they frequently come right up to my face, blowing bubbles at me. Their behavior can only be described as curious and playful. Back in the boat I watch them surfing waves near the shore, their bodies silhouetted as the waves rise, flippers and

3.8. Sea lion silhouette

bodies outstretched. I am convinced they are not only playful and curious but absolutely blissful in the water and with each other.

After seeing the formation of spotted eagle rays, the surfing sea lions, and the machine-like conveyance of birds dive-bombing fish, it occurs to me that even our most cutting-edge technologies and inventions pale in comparison to the mother of all invention and technology and evolution itself, Natural Selection. Her designs are honed through millions of years. How can we expect to better them? In designing anything, we should first look to nature for advice. The birds we watch are specialized for aquatic assault, with pointy noses and tails to streamline their buoyant torpedo-shaped bodies, air sacs in their skulls to cushion the blow into the water, and nostrils that close to keep water out. The most sophisticated billion-dollar war technology cannot compare. As a human, I feel like a soft, vulnerable, humble blob in comparison to what Natural Selection has crafted.

The Red-lipped Batfish—Lumpy, Bumpy Apparition of the Deep

Near the end of our trip, we are scheduled to make a night dive to search for the red-lipped batfish. After seeing what's down there, we are a little bit nervous about diving into the deep dark at a time when sharks normally hunt. After dinner, we struggle to pull on wetsuits that are still wet (and now very cold) from the morning dive and stand shivering on deck to get our instructions. We fidget with the huge heavy flashlights we each carry, wondering how we will check our air gauges or fiddle with our masks with both hands full. We are going to slowly descend, holding hands so we don't lose sight of each other, to about one hundred feet, at which point we will navigate along the sandy ocean floor until we encounter the small and funny-looking batfish. Sounds simple enough, I think: straightforward, predictable, and safe. I'm ready. We line up one by one and jump in, feet first. It's a good thing that each person behind can't hear the muffled screams of the person ahead; otherwise we would never all have gotten into the water. As we hit the water, it feels like icicles stabbing into every inch of our bodies, and we gasp into our regulators and emit garbled sounds, which doesn't feel safe at all. The next shock is the pitch-black abyss we are looking down into. It was bad enough seeing the sharks below us, but imagining them all around without being able to see them is even worse. We all turn on our flashlights in a hurry, scanning the area around us, expecting to see teeth. I look at my heavy light with new appreciation as I note that it can be a bonking-on-the-nose device. No teeth in sight, but I know they're out there, waiting like piranhas just beyond the scope of the protective force field of light.

I yelp as the bottom comes up and hits my foot. Are we already one hundred feet down—on the bottom? As soon as we spot rocks, sand, and plants, everything changes. We have something to focus on, something to "ground" us, literally. I realize at that moment what land creatures we truly are, that being on land is embedded in our very genes. We are no more adapted to hang in dark open water than a rhino is, and it is expressed in our every thought and action. The choices we make are not our choices, but the choices of our genes directing us—to stay near land, to live in a house, to reproduce. It may thrill us to play with our fear of foreign things, but we never stray too far from what is best for our survival.

After plodding along the bottom for some time, quite comfortable now, with our noses to the ground, we spot our first batfish. The sight of it inspires an involuntary laugh, which sends a fit of bubbles through my regulator. It is the most ridiculous thing I've ever seen! I watch, fascinated, as this miniature mythical beast wobbles away in front of us. I barely have to move to keep up with it. Kicking lazily along above it, I can see right away how it got its name. Bony front fins stick out from a skinny warty body at a rakish angle, helping it walk and flutter along the bottom. Its broad head has a pinched face and an unhappy-looking mouth covered in white whiskery-looking protrusions that flutter in the current. It looks like an angry drowned bat, and is no more graceful than a bat would be, flopping around in the cold dark water. Strange

3.9. Walking batfish

only begins to describe this would-be fish that looks like its evolution was arrested between frog and goblin. In fact, it is a relative of a group of fishes called the frogfishes. It hobbles along the sand and rubble of the ocean bottom on prehistoric-looking leg-fins, and only when its life is threatened does it attempt a herky-jerky version of swimming that reminds me of Frankenstein. On its wide alien head, I kid you not, is a big fleshy horn that sticks straight forward, like it's jousting with a moldy sponge for a sword. The horn hides a lure thought to attract prey, but not much is known about how the fish uses it. I guess it needs a ruse to catch the worms, crabs, and small fish it eats, since I can't imagine it chasing anything down. To top off this ridiculous ensemble, the grumpy mouth is framed in bright red lipstick.

Well, at least I don't feel so ungainly now. If that creep from the deep can live down here, so can I. Our air and time runs out quickly at this depth, and soon it's time to go back up. Looking down at the batfish, small and ugly and alone on the cold ocean floor, I am overtaken by a wave of sympathy. Like me, it's awkward in the water and vulnerable to the ocean's real machines. I have an urge to take it with me to a world where its feet at least make sense. But of course I can't—the ocean bottom is its home. Who knows how a batfish feels? Perhaps it just accepts. In any case, seeing the batfish made a lasting impression on me—the thought of that creature living at the bottom of the ocean while we drop off our dry cleaning and paint our nails. Who is more ridiculous, in the end?

———————————

The batfish somehow leaves me lighthearted and optimistic about the Galapagos Islands. Despite the human-caused destruction playing out, there remains cause for hope. After all, no species have been known to go extinct since the area was declared a national park in 1959. In fact, two species thought to be extinct have been rediscovered, and some brand-new ones have also been discovered. Several species are threatened, but many populations have been restored, most notably the giant land tortoise, the land iguana, and the Hawaiian petrel. We never were able to find the Blackspot chromis, which could be meaningful as a portent of future climate change. We did find plenty of sharks, and great progress has been made recently to prevent shark finning. In July 2007, the Galapagos National Park Service began to crack down on illegal shark hunting by patrolling with planes and boats in the northern Galapagos Islands where most of the sharks are found. In addition, dogs trained to sniff out shark fins have been put to work at the airport, and stiffer legal penalties have been imposed on lawbreakers. Most important, each one of us, in concert with great ambassadors like Jean-Michel Cousteau, Dick Murphy, Jack Grove, and others, became an advocate for this beautiful and wondrous marine world and joined forces in the fight to protect it.

FOUR

Chasing Nightly Marvels in the Rocky Mountains

Rick A. Adams

Encounters with a Night Druid

Dusk. Kate and I are hiking into the eastern foothills of the South-ern Rocky Mountains. As shadows lengthen, the woods begin to come alive with animals accustomed to working the night shift. A towering thunderhead lights up and bellows above, momentarily spinning dusk back to familiar day. But darkness is winning the hour. From a distance, a nocturnal bird awakes from its diurnal slumber and calls out its name into the night air: *poor-will, poor-will, poor-will.*

Although we have hiked on this trail on many evenings, I admire its beauty as if it were the first time. The trail parallels a small stream and is rich with singing birds and beautifully colored insects. As the summer has progressed, we are in excellent shape, and traversing several miles of steep terrain to set up our bat nets in the evening has become routine. We head into the foothills while people out for an evening stroll are headed back to the comfortably well-lit city. I love evening because the night predators begin to stir, and we occasionally observe bears, coyotes,

DOI: 10.5876/9781607322702:c04

red foxes, owls and, on very rare occasions, the enigmatic cougar, heading out to forage. The sky darkens as we continue up the trail.

Working at night in wild places changes a person. One becomes more and more integrated, and the typical trepidation of wandering through the woods at night becomes a long-lost memory. But tonight will bring a very different kind of exposure, the type that will challenge the senses of even an experienced night wanderer. In fact, the only reason I know for sure it happened or that I venture to tell this story is because there was another witness to these events.

Kate, a student from Boston College, has been working with me on the ecology and behavior of the nine common bat species distributed throughout the eastern foothills of Colorado. She is resilient, competent, determined, and has an open but stable intellect: an important attribute for interpreting what is to happen tonight. Because we got a bit of a late start, we are hiking quickly to get to our netting site before the bats emerge. We are making good progress.

We have seen no one else on the trail, so when a human figure suddenly materializes from around a bend ahead of us, just at the edge of my twilight vision, my focus is immediately drawn. At first, it is hard to tell if the silhouette is coming toward us or moving away. After a few more steps, I surmise that she or he is headed in our direction. In another minute, I realize that the distance between us and the figure is closing rapidly, and for no apparent reason, warning signals begin resounding in my brain. Draped in the darkness of a black cloak that appears to absorb rather than reflect light, the human-like image, bolstered by a long cane with every step, moves toward us with an exotic but resolute gait, causing an unsettled feeling to overwhelm me. As our paths cross, time seems in slow motion. I hear no sound nor sense a life force; rather, its essence is like cold stone fingering and pulling at my passing vitality. The fleeting encounter lasts only an instant, but the convicted embrace of that moment hangs with me and my unease is overcome only by refocusing my attention on getting to our field site on time. I say nothing to Kate.

4.1. Author taking field notes on a long-eared bat

When we reach our destination, we begin to deploy our mist nets. The nets are set between two poles, not unlike those for a volleyball game. But mist nets are finely crafted to be "invisible" to birds during the day and imperceptible to bats, which use echolocation at night. Despite its subtleness, the mesh, made of tough nylon, is quite durable. The nets are designed with pockets that billow and gently slow a bat as it hits and then slips into a hammock-like sling that is usually inescapable.

Our studies concern the abundance and distribution of bat species in foothill habitats and resource use, particularly water. Over the following hours we catch several bats—a few little brown myotis, a big brown bat, a fringed myotis, and two long-eared myotis. We carefully measure and weigh each individual as well as note its reproductive condition and general health before releasing it. On a few lactating females, we place a small radio transmitter that allows us to follow the bat using a telemetry receiver that captures the signal, so we can track her foraging patterns and also locate her daytime roosting site.

We have been catching bats for several hours now, and as activity wanes, we complete our release and finalize our field notes. We break down our gear, pack it up, and descend the canyon, somewhat resistant to returning to the chaotic city. It is midnight. Only a partial moon lights our way. We turn off our headlamps and relinquish ourselves to the darkness, walking as our ancient kin once did among unlit shadows. As we retrace our steps down the trail, we round a bend and approach another of the shadows caused by the lining foliage. However, I soon realize that another world awaits us inside this particular shadowy nook. Unexpectedly, time slows as a transparent sketch of cane and man steadily takes form within the shadows. Hairs on the back of my neck pull at their roots. The familiar sounds of night evaporate in this uncharted dimension; cold hollowness fingers me as I pass. My first response is to flee, but as a scientist I need to know. I start to avert my eyes, but an unrelenting urge to look deep into the shadows of this unfamiliar world overcomes me. But the deeper I look, the more transparent the figure becomes, eventually dissipating completely into darkness. During this time I am unaware of Kate and what she might be experiencing. As I leave the shadows I feel relief and regain my senses.

Our descent has been and continues to be undertaken in silence, which is typical after a long night's work. As we distance ourselves from the shadows, my concerns give way to self-reassurance via denial. In my own head I become convinced that the episode was a self-conjured delusion of a confused night wanderer; perhaps it's time for me to seek a day job, I think. Suddenly, my thoughts are broken by Kate's voice: "Did you feel as though there was someone or something standing in the shadows on the trail back there?"

Nights at 12,000 Feet

As a yearning biologist enrolled at the University of Colorado, Boulder, I was hired in the summer of 1986 by a well-known ecologist to work at the Niwot Ridge Biosphere Reserve, run by the university's Mountain Research Station and Institute of Arctic and Alpine Ecology. The research area is located at 12,500 feet, about 1,000 feet above tree line, where it can snow on any day of the year and winds can be relentless. The warning signs of an approaching harsh environment as one nears tree line are clear and embodied in the grotesque forms of recumbent conifer trees stretched across the landscape, emulating dying foot soldiers. These vegetative growths, known as *krummholtz* (German for "crawling"), are emaciated similes of trees that grow to thirty feet in height a mere 1,500 feet downslope. Deformed and sometimes freakish with shortened twisted limbs or flag-poled trunks, the "soldiers" shout to all comers: *Turn back!* The tree-line environment, where forest gives way to seemingly barren tundra, is one of the most dramatic transitions between ecosystems in the world. Here, krummholtz literally crawl slowly but relentlessly across the tundra inch by inch year after year, their dying branches extending downwind as new upwind growth fingers forward.

But even the hardy krummholtz cannot eek out a living where I am headed. Once above 12,000 feet, cushion plants rule, and life itself becomes diminutive, a scaled-down microcosm of biodiversity. Most of the food for herbivorous animals on the tundra exists belowground. So many animals in this environment are small and spend most of their lives burrowing or scurrying through habitat created under the rocks of talus slopes or through the soil. Amazingly, birds persist in this two-dimensional treeless landscape. Building nests on the ground, leaving them and their eggs quite open to predation, but somehow beating the odds. One bird, the ptarmigan, is an alpine specialist that survives above tree line even throughout harsh winters by morphing between brown and white colorations that blend with the seasons.

It is summer in the alpine, and I have been hired to trap small mammals, including mice, voles, shrews, chipmunks, pikas (small alpine rabbits), and marmots, for a yearly population census. Most of the plants around me are no taller than an inch, and a cold unobstructed

4.2. August snowstorm at dawn, Niwot Ridge, Colorado

wind from the northwest hits my bare face. Not twenty miles away, but 6,000 feet down, the air temperature is a sweltering 100°. Clearly, I have traveled much farther environmentally and climatically than I have physically. An arctic feel caresses this place.

Carrying gear and supplies to last ten days, I stop every one hundred feet to regain my breath. After about two hours, I reach my destination and look forward to the relief of dropping my pack. The accommodations consist of a 1940s retired army trailer that was pulled by a tractor up onto the tundra decades ago. At times, six or seven biologists share the very small trailer full of field gear, which holds four bunk beds with mattresses that should long ago have been discarded or perhaps burned. Residents on this trip are Richard, a master's degree student working on pikas, and Mike, an undergraduate hired to help me trap and tag anything we catch. We are a cramped but happy bunch as we are doing what we love, and the Ridge (as we call it) is a haven from the hectic life and pressures of graduate school and the city below.

4.3. Saddle Van, 12,000 feet, Niwot Ridge, Colorado

We trap for pikas and marmots during the day, but the mice, voles, and shrews are night lovers. Consequently, we open little mouse hotels (Sherman live traps) in late evening and check them at dawn. It is August 12, 1986, three days into our stay, and we have just opened and baited 300 traps set in several grids on the tundra. This is backbreaking work, and it takes us three or more hours, this coming after trapping and tagging pikas and marmots since dawn. Finally, at 7:00 p.m., we are ending a long day that began at 4:00 a.m.

As we settle in to eat dinner, we hear an ominous sound. A thunderstorm is rumbling across the western sky. Luckily, we are in our tin shelter. One of the biggest concerns for biologists working in this environment is being struck by lightning, but it is not only our safety that concerns us. The pikas and marmots we capture are in open-screen traps—exposed to the elements and unable to survive a storm. If a storm approaches (and here storms come in quickly), we must check all the traps and release the animals before heading to safety ourselves. On

4.4. Author setting Sherman live traps, Niwot Ridge, Colorado, 1987

one occasion, as a storm encroached, I was closing the thirty or so traps as fast as possible. I came to the last of a series of marmot traps—one that had not caught a single animal previously—and saw a fat and not very happy *Marmota flaviventris* moving nervously in its temporary prison. I ran the last few yards to the trap and began to process the animal as quickly as possible, coaxing it from its cage into a squeeze trap, which is basically a funnel of cloth with metal bars in a cone figuration at one end. As the animal moves through the sack, it is squeezed by the cloth into the metal cone where only its head protrudes, allowing the biologist to ear tag, weigh, sex, and release the animal while avoiding injury to both. In about ten minutes, the marmot was processed and gone, but the storm had moved in directly above me as I jumped up and began heading to the protection of the trailer.

As I approached within twenty yards of safety, a cloud discharging virga (rain that never hits the ground) loomed ominously above me. I began to panic when suddenly my hair stood straight up on end,

4.5. Yellow-bellied marmot, Niwot Ridge, Colorado

crackling with electricity, a sure sign of imminent electrocution—and death. A lightning bolt extended downward, but rather than striking the ground or me, it seared across the sky a few feet above me as thunder and the accompanying expansion of air threw me to the ground. The bolt struck the top of a nearby ridge. Dazed, I began to crawl on hands and knees across the tundra toward the trailer, but my progress was too slow—I jumped to my feet and ran to the safety of the trailer.

But now we are all in our safe haven, although at times sheltering in the metal trailer feels like being in an oven awaiting a lightning strike, and within a few minutes our relative ease turns to anxiety. It begins with a *tick, tick, tick* on the metal roof of the trailer, followed by an eruption of cascading hail of such intensity the trailer begins to rock and shake violently. We cover our ears. The sound of ice striking metal at high velocity is deafening. Although the ice storm is over in a few minutes, the damage is more than done. We know that all the spring-loaded live traps we so carefully set to trap mice will be tripped shut by the impact of the hailstones. In addition, upon opening the door to the trailer we realize they will also be buried in an inch of ice cubes and the temperature has dropped below 40°. So much for dinner! There is no time to be lost. We head back out to work, hungry and grumpy, to reopen all the traps.

As darkness creeps in, we finish our work by flashlight. The sky is turning a lacquered array of colors, accentuated by the lack of atmosphere. The air is cool and crisp, and the Milky Way is visible in ways not observable from lower elevations. The stars are overwhelming in number and brightness. On this moonless night, the extravagance of the heavens is eye candy, making it almost impossible to look away. Suddenly a meteor streaks across the sky from horizon to horizon, and I remember that we are in a time of year for the Perseid meteor shower. The night comes in cold and windy, but we manage to stay out and watch some of the show as rocks turn to fire and light up the sky from all angles and dimensions, "raining fire in the sky," as John Denver described.

We return to the trailer at about 10:30 p.m., eat a bit of food, and retire to our bunks. Lights out and all is quiet except for the wind that occasionally intrudes into our consciousness, but we are too tired to care. We are dead asleep at 3:00 a.m. when a visitor arrives beneath the trailer and awakens us with its labored movements, accompanied by a series of muffled bangs and grunts. At the edge of consciousness, I begin to question what is happening, but stillness quickly returns and I again drift off. Moments later and with the grand entry of a Broadway show, a deep wave of coarse sounds floods the trailer in concert with earthquake-level vibrations. Quickly it becomes clear that something is trying to take bites out of the trailer. We stumble from our bunks, running into each other trying to find a flashlight. Richard opens the door, drops to his stomach, and hangs down, shining the light under the trailer, and after some searching, locates the beast near one of the rusty metal wheels on which some remnants of rubber tire hang.

Our visitor is indeed one of the strangest animals in North America. It is an eating machine with huge, ever-growing incisors, and its body is covered in modified hairs that give it protection from all perpetrators intending harm. They amble away their days, usually eating wood. But this one apparently has taken to metal and rubber. As you have probably surmised, our surprise guest is a porcupine. But here at 12,500 feet! There is no natural wood to be had in this treeless environment more than 1,000 feet above tree line. What the hell is going on? This apparently deranged porcupine is delightedly gnawing on the remainder of the rubber tires and metal wheels under our forty-year-old trailer. Discovered by our flashlight, it suddenly stops, and silence returns as it moves away from the light. We wait about five minutes, and as the silence continues, we assume—hope, really—that it has moved off and peace has returned to the night.

As we begin to slumber once again, comforted by the thought that this bewildered visitor must have hiked back to where it belongs, feeding properly as nature intended, the racket suddenly resumes—this time, however, with much more intensity. We bang on the floor of the trailer as if we were living above an apartment full of noisy teenagers partying. Just like teenagers, our new friend ignores our not-so-subtle hint. At this point, bewilderment becomes contempt, and Richard behaves as unpredictably as a porcupine hiking around at 12,000 feet. He jumps from his bed and leaps out of the trailer—a "caveman" wearing white underwear and hiking boots without socks. He grabs a seven-foot-long wooden pole used to measure snow depth in winter, dives headfirst under the trailer, and proceeds to try to dislodge the unwelcome guest. But porcupines are hardy and innovative, and this one is using its quills to wedge itself in a crevice under the trailer, warding off attempts to dislodge it. As the wind gusts and whirls, Richard is grunting, cursing, and tussling under the trailer. Mike and I watch from the doorway in amazement. Richard's legs stick out from under the trailer, contorting and flailing. For what seems an eternity, intermingled yelling and cursing from Richard along with whining shrieks from the porcupine spin through the frigid night air. Finally, Richard is successful. He dislodges the porcupine, and it takes off running across the tundra. But this is not the end of the saga. An adrenalin-pumped Richard emerges from under the trailer and runs in his underwear in subfreezing wind chill across the tundra, chasing a porcupine fleeing for its life. Pursuer and pursued disappear into the darkness as meteors continue to rain down from the heavens.

After a few minutes Richard reappears from the darkness with a big grin on his face, sure of his success and proud of his victory. We laugh and congratulate him for his determined efforts. We return to our beds and to our slumber—the night, what is left of it, is now ours. I doze off to take advantage of the last hour of darkness. I have just begun to dream when the deafening sound of gnawing from beneath the trailer grips the night once more.

Cougars at the Nightstand

Most mammals are nocturnal, and this certainly goes for the charismatic megapredators that intrigue so many people. Throughout the Rockies, large-bodied predators such as black bears and mountain lions prowl the night shadows. For the bears, an omnivorous diet consisting mostly of vegetable matter makes up the menu. Mountain lions, however, adhere strictly to the Atkins Diet, eating very little besides meat, which they acquire by ambushing their prey (usually deer), and darkness provides perfect cover for such a hunting strategy. The victims likely never see their attacker; the strike comes from above and behind with no warning, crushing the back of the skull.

My encounters with mountain lions have been many and varied. In some cases, individuals have sounded their presence by screaming a territorial plea apparently meant to move us out of their province. In most cases, if the plea goes unheeded, the lion gives up its demand and moves away. However, in one particular case, a test of wills ensued.

It is July 1999. Krista, a graduate student, and I are netting bats at a small water hole in the ponderosa pine woodlands of the Rockies. On the hike in, Krista expresses how much she would love to see a "big cat"

4.6. Mountain lion

this summer. Aloud, I tell Krista that sightings happen occasionally. My unspoken response is: *Be careful what you wish for.*

As darkness approaches and the bats begin to fly, a familiar and unsettling sound is aired, as if conjured by Krista's wish. It is indeed a cougar, and it is screaming at us to leave the area. I tell Krista not to be alarmed, that it will most likely give up and move off. However, instead of retreating, the screams persist, and the cat seems to move closer over the next thirty minutes. The nearer the cat comes, the more desperate its cries become. Forty minutes pass and the only change is the level of vocal desperation coming from the cougar, now including deeply guttural and mournful moans and groans. I seriously consider leaving the area, which would be a first for me. Krista and I stand in silence. I shine my headlamp in a 360-degree circle to try to spot the beast. In an instant, the lion's eyes appear from atop a rock on the adjacent slope, a mere one hundred feet away. The cat peers intently, blinking momentarily as it moves from its perch in our direction. I begin to panic somewhat, but not in a way that Krista would notice, and I tell her in a calm voice that we should pack up and leave. We break down the net, nervously and hastily cramming the gear into our backpacks with little care. As I rise from bended knee, my headlamp strikes yellow eyes shining back at me, now from twenty feet away. I can see the entire animal. The cougar's ears are pinned back and its stance is erect and rigid, a clear threat display. I reach down and scrabble the ground for a small rock, which I toss elliptically to the side of, but near, the cat. The cat averts its gaze from us, looking in the direction of where the rock hit the ground. As it is momentarily distracted, we move quickly, hiking down and away in silence, our actions apparently appeasing the cat, as we hear no other sounds from it as we scurry away.

As we descend, I begin to wonder why this event happened. What was it about our presence that stimulated the persistent demands for our departure? Was the cougar trying to access water? Probably not. There are many areas along that steam where a cat could drink. Were we near a cached kill and was the cougar was coming back to feed? Possibly. Cougars cache large prey (a deer, for example) and revisit the carcass to eat for days.

The next day, I discussed our adventure with the local rangers, and two of them hiked to the site to determine if a cache was indeed present. None was found (however, such caches it can be difficult to find). Another explanation for the encounter might be related to reproduction. From this individual's size and slender appearance, it appeared to be female. Perhaps it had young kits nearby and saw us as a threat to their safety. Whatever the case, a lesson in persistence was learned—by us and by the cat—this night.

Although that was the first and only time a cougar actually approached with the intent to move me from an area, other encounters with these large cats have been equally unnerving. One was years earlier. On this occasion, Kate and I are catching bats at a small pool of water that forms as a trickling stream crosses a footpath. The footpath makes an easy access "route" for flying bats to descend and skim the surface of the pool to drink from a cool mountain stream. We string mist nets across small water sources like this because bats concentrate around them and they are not expecting our nets. Otherwise it is almost impossible to catch bats in free flight; they are simply too fast and maneuverable. Tonight, a full moon greets us and at once highlights the landscape's rugged beauty while exposing mysterious shadows and ghostly shapes that evade focus, blending imagination and certainty. "Business" is slow as it is a rather cool night and our winged friends apparently are not seeking a cool drink at the stream. In silence and half asleep, Kate and I sit on the ground next to our nets, waiting patiently. We know this is not going to be a record night for numbers of captures, but it is sometimes on slow nights like these that a rare species, such as an eastern red bat or even a Mexican free-tailed bat, shows up.

As I gaze, half alert, into the darkness at the opposite edge of the small pool, I notice a slight blur of motion that breaks the edge of shadow and moonlight. For some reason my brain signals that something is amiss. I am in mid-thought when Kate whispers, with equal hesitation and concern, "What was that?" I immediately trigger my head torch, only to be astonished at what stands before us. A mere six feet away (remember, we are sitting on the ground) are two fully grown mountain lions, and we suddenly find ourselves literally face-to-face with the region's apex predator, which can weigh 165 pounds and is known to have a pouncing distance of thirty feet. We are sitting ducks. As the light from my headlamp hits them, fortunately the cats are momentarily confused, stretching their necks as if trying to peer through a sudden and unexpected sun.

A moment later, I find myself standing and instinctively yelling while breaking into some sort of primal dance, throwing rocks and sticks in an attempt to intimidate the predators. Kate is also standing, but stunned and motionless. My dog, Jasper, a husky-golden mixture, is sitting in his usual calm state, watching all this with restrained curiosity, almost amusement. The cougars respond to my actions by backing off about ten feet, but hold their ground. I yell louder and more emphatically. I shout at Kate to hand me rocks. It would not be wise for both of us to bend down. I throw several large rocks, not really trying to hit the cats, only hoping to scare them off. Eventually the larger of the two takes off down the trail, apparently having had enough of our antics. The smaller lion, however, cannot still its curiosity and begins to crouch and move behind the vegetation in a circling pattern. I continue to throw rocks and sticks, but the situation lacks improvement. My adrenalin peaks. I reach down and grab a baseball-sized rock. Winding up like Sandy Koufax, I hurl my best fastball. It meets its mark, pasting the remaining lion on its side. The cat lets loose a resounding groan and runs down the trail after its buddy. I release a primeval scream from depths previously unknown to me. Kate and I breathe sighs of relief and decide to pull out the nets and leave the site.

As we descend the trail with a forty-five-minute hike ahead of us to civilization, my mind begins to wander into thoughts of deep time, the long distant past when our species—and, before them, other hominids—were commonly faced with threats similar to and surely more terrifying than this one. Then the stakes were much higher. Prowling carnivores were more diverse, and many were gigantic by today's standards. For example, in North America, scimitar cats the size of African lions stalked the night, as did two species of saber-toothed cats, the largest of which, *Smilodon*, weighed up to 650 pounds. The short-faced bear, which dwarfed today's grizzly bears, roamed at will. An American cheetah, *Miracinonyx*, coevolved in North America with pronghorn antelope, thereby selecting for their high-speed running ability, which far outpaced any current predators on the range. There were, of course, mammoths and mastodons and bison that were six feet at the shoulder and weighed more than 2,000 pounds. Even beavers reached gigantic sizes of 600 pounds or more.

Those were, indeed, different times, and my thoughts of them brought our encounter with the curious cougars into perspective. Past humans had to contend with daily threats beyond our imagination, and they did not have a house or car to retreat to. It is humbling to think about the path of human evolution compared to the sterile surroundings many people prefer today, living an existence almost entirely separate from, and fearful of, nature. I never have felt more alive than I did just after this incident, an obvious side effect of adrenalin. The cats, likely young siblings that had recently left their mother, clearly were curious rather than aggressive. Otherwise they would have pounced on us before we even knew they were there.

Year of the Bears

It is 2001. The regional drought is in its fifth year. Little snowfall in the Rocky Mountains over winter creates the impression of an old worn landscape that can barely support life struggling to reemerge from its wintry slumber. Regrettably, spring regeneration has become an ancient memory for much of the regional plant life. In May I hike to the 11,000-foot pass of Arapahoe Peaks, hardly traversing any snow at all. I am wearing hiking boots when typically snowshoes would be required for much of this trail for at least another month. Cascading waterfalls that usually drown out nature's other sounds are mere trickles of tears from mountain slopes that appear to be dying. In the foothills, little growth or renewal is apparent. Mountain streams, if flowing at all, are remnants of past discharges. Forests are ready to burn—and burn hot. Warnings of the severity of the crisis are posted as early as May; total fire bans are in place before Memorial Day.

This drought will have immeasurable effects on almost all wildlife. But predicted to suffer the worst are the large omnivores such as bears, which require huge amounts of plant food not only to survive summer but to put on fat for the lean winter months beyond. Black bears (*Ursus americanus*) are not uncommon at my field sites. It is estimated that 3,000 individuals live along the Front Range between Colorado Springs and Fort Collins. Although present in high numbers, they are rarely observed by humans. There are only a few sightings a year, usually individual bears that wander into the cities near the foothills. But this year is different. The plants that the bears rely on to sprout berries in August, at a time when foraging consumes as much as twenty hours of a grown bear's daily routine, are producing their fruits in early June instead, hedging their bets against the oncoming and increasing drought conditions. The plants are pulling out before it is too late, cashing in their reproductive currency before the moisture runs out completely. When August rolls around, the bears will have no food when they most need and expect it.

The impending hunger will predictably drive some bears into the cities looking for food in dumpsters, trashcans, pet food bowls, and even hummingbird feeders. Colorado has established a policy of three-strikes-and-you're-out, meaning that the third "infraction" is literally the last for that bear. Caught for the first time in the city, a bear is hazed back into the forest or tranquilized and moved. The second time that individual enters the city, it is tagged and relocated to some far-distant place (usually 100 to 200 miles away). If it returns, the third strike is a death sentence. One of the most amazing illustrations of the mobility and determination of black bears to remain at "home" occurred during a drought. A male was captured a second time in the city of Boulder and was subsequently moved to Alamosa, Colorado, about 250 miles away in the south-central part of the state. Within a month, the bear was back, having hiked the distance across unfamiliar landscapes to get home. Much of the problem associated with bears entering the city and coming in contact with humans is caused by people's irresponsible neglect when it comes to securely disposing of their garbage, loading bird feeders with fatty seeds, and leaving bowls of pet food outside. If these temptations are removed, bears generally stay away from humans and towns.

By August, as predicted, the foothill bear population is suffering, and there is little natural food to alleviate the bears' hunger. As cruel as this seems, it is the way of nature and natural selection on animal populations, controlling numbers and density through space and time.

4.7. American black bear

During this summer, I frequently encounter bears roaming widely to find food when I'm out conducting fieldwork at night. Since many of these encounters occur in complete darkness, I'm aware of my companions not through sight, but through smell. I rarely hike with my headlamp on at night. Part of this is to save battery life, but mostly it's because once one's eyes adjust to the darkness, seeing and navigating the trail is quite easy. There are many occasions when the strong musky smell of a bear makes clear its presence; it had probably crossed or left the trail as I approached.

On an evening in August, Christina and I are hiking through one of the most beautiful foothills trails in Boulder County to access a small water hole where we will net bats. We follow a small stream, noting the area is full of birds—western towhees and mountain chickadees—nesting close to some of the only available water throughout the area. The trail sings with life, even in a year of so much death. We are moving through a riparian zone as dusk creeps in. The landscape opens up a bit into a montane meadow, where just across the drainage we see a black bear rummaging among the skunkbush for food. It appears, by body size, to be a young male, and it has a white stripe down its chest. It notices us and becomes motionless, then sits back on its haunches and begins watching us. We too become motionless, watching for a while through binoculars. As we begin moving away, the curious bear begins to forage again.

Daylight is fading now, so we must quicken our pace to get to the water hole in time to set our nets. After a very steep incline, the trail levels off. We round a bend, and in the middle of the path is a large male black bear standing in profile. We approach a little closer, but the bear does not budge. I decide that it's best if Christina waits while I walk ahead in an attempt to move the bear off the trail. Usually, bears will move away when a human draws near, so I am confident this will not be a problem. As I walk toward the bear, yelling, there is no response. Ah, stubborn, eh? Not to be deterred, I keep encroaching on the bear, clanking together the bat net poles I am carrying. Surely this loud and sudden noise will dislodge our roadblock. This particular bear, however, would have none of it. Until this very moment, the bear had avoided eye contact with me, looking off-trail in the direction I was trying to move it. As soon as the clang of poles rings out, the bear turns its head, glares at me with obvious contempt, and begins lumbering down the trail directly at me. Yikes! I stop in my tracks and begin to backpedal as fast as I can, not wanting to turn my back to the bear. Once I reach Christina, we both slip around the corner out of the bear's sight. At this point, Christina begins laughing at me and I don't blame her. All right, all right, I'm not as brave, or stupid, as I pretend. We back up a bit farther down the trail, expecting the bear to come around the bend, but it does not. After a couple of minutes with no bear, I decide to check on its whereabouts, walking ahead and peering around the corner. By now it is almost completely dark and I turn on my headlamp. Hoping for the best, I peek around the corner and see our big furry friend still standing in the trail, unwilling to concede right of way. I look up and see bats flying everywhere, foraging up and down above us. After all this, time and the recalcitrant bear win, as it is too late for us to set nets.

We start our long hike back. Occasionally on the way down I grumble about that damn bear, hell-bent on disrupting our field study. When we reach the trail junction near the spot we saw the first bear in the riparian area, it is now too dark to see it; but we can hear and smell it (or some other bear) foraging down in the same drainage, right where the trail moves through. The sounds of cracking branches and the turning of rocks echo up the trail. It is now completely dark, and we decide to head down on another longer route to avoid any further bear encounters. A musky smell fills the air, reminding us that we are merely visitors in another's home.

The Curious Fox

As I explained above, we net bats at water holes because at these areas bat activity is concentrated in a confined space. Most bats must drink nightly, and thus eventually most bats in the area will show up. This, of course, is true not only for our bats—water pools draw in many more animal species.

Many species of insects breed in water, and some have aquatic young with gills that metamorphose into a completely different form as an adult. In fact, the adult form is usually some aerial version that in many ways does not resemble the aquatic larval stage at all. Although even children are cognizant of the complete metamorphosis of moths and butterflies, fewer people know about the hidden lives of aquatic-emergent insects such as dragonflies. Dragonflies are aerial predators as adults, eating other insects that they capture while in flight. I have witnessed huge hatches of dragonflies around ponds at my field sites, at times so many flittering about at once that the sky appears filled with huge eyes and transparent wings. As adults, dragonflies are certainly one of the oddest-looking animals on earth. However, their hidden life among the underwater rocks of ponds and streams is equally astounding. Voracious predators, they manifest large protruding mandibles that function as propellant spears that they thrust into their prey so quickly that the human eye can barely perceive this violent motion at all. As a body form, dragonfly larvae appear alien, and if such an organism was several feet long as apposed to an inch or so, few people would venture to swim in lakes and ponds. They are truly terrifying to look at. But it is the aerial version that dominates terrestrial landscapes, and as I see tonight as we pass a small pond on our way to our netting site, they can occur in overwhelming numbers. Looking at the scene, my thoughts drift to the Cretaceous period (70 million years ago) when dragonflies had wingspans of three feet and eyes as large as dinner plates. Imagining how this scene would have looked then is rather overwhelming and not for the faint of heart.

As darkness dims our visual world, other animals with eyes adapted for night ramblings are on the move, leaving the resting places that protect them from the daytime heat. Under the cloak of night, mammals with big brains thrive. Indeed, another visitor to the water hole will provide us a glimpse into one of these wild and complex brains. Near this particular site, we are aware of a red fox den located under a large rock in a heavily wooded area of Douglas fir habitat. The den is close enough to the trail that a person hiking at dusk might witness the enthusiasm of newborn life in the form of fox pups frolicking about in careless play. I have counted up to four of them, peering with caution from the entry to their den, trying to assess the threat of our approach and no doubt learning about what makes a human.

Mammals have the largest brain per body size of any organism yet evolved on Earth, and the evolution of mammary glands for feeding their young nutritious milk has led to significant levels of parental care for many mammal species. Such intensive contact for extended periods of developmental time gives mothers ample opportunity to teach their young how to survive. All of this has apparently led to the evolution of a brain that is more plastic, or open to learning, as opposed to one built prevailingly upon purely basic instinct. The deep recesses of the convoluted mammalian brain provide plenty of surface area and integration of neurons for complex learning and memory storage. Mammals are clever, manipulative, and sneaky. With about 5,200 extant species, they cover the globe. Their evolutionary radiation has led to a worldwide distribution of "environmental manipulators" of which even the most ancestral (primitive) species manifests uniquely mammalian characteristics for survival.

I have had the pleasure of watching the behavior of various bird species and tree squirrels that use a water bowl placed on the deck of our house. In many cases I am sitting in full view and only a few feet from the bowl. A bird looking to quench its thirst typically lands on the bowl, and if I make a sudden move, the bird flies off rapidly, disappearing into the sky. In contrast, the tree squirrel has a fundamentally different response to the same threat. It may initially retreat if I suddenly move, but it will not usually leave. Instead it moves about the deck or just off the deck in the grass, observing me and testing its chances of success for getting a drink of water without being eaten. Although clearly wary, it appears that it cannot contain its curiosity. It moves to different areas at different distances (closer, then farther, back and forth, back and forth) and watches my response, learning about me. If

I sit silently without moving, the squirrel, although wary of me, will inch closer to its goal until it determines either that I am not as much of a threat as originally thought (too slow to catch it), in which case it comes to drink, or that it should not approach—and it sprints off to the nearest tree.

A close observer can actually note the squirrel going through the mental gymnastics of problem solving and accumulating information to make this decision. This is not to say that there are no smart birds out there. Quite the contrary, but the sneakiness and determination that a mammal uses to solve problems appear to be unique.

And as we spend another night at a small pool of water under the stars, we are visited by one of the more cunning mammals on earth. A red fox, which by all rights should be out hunting for food. But this particular individual is more interested in learning, or at least observing, something in the environment it apparently had not before seen: two bat biologists doing their thing. Exactly how long the young fox has been watching us before we are aware of it, I don't know, but it has certainly been watching us *closely* for at least an hour. I notice its eyeshine only after it perches itself in the open on a large rock about twenty feet from the water hole. This has been an active night, and we are busy carefully removing bats from the net. I would love to record this animal's reactions, but time does not permit such an extravagance. All I can say for sure is that this fox sits in the dark for more than an hour, attentively watching us as we capture, mark, and release bats.

What about this could entertain this wild mind for so long? And what is it learning? Although its attentiveness might be interpreted as that of a simple mind, entertained by simple acts, we biologists know better. The carnivore mind is one of the most complex and is developed around memory and learning. Predators must be able to alter their behaviors to accommodate changes in prey populations and environmental conditions and must be able to compete with other smart carnivores. No, I don't think this is a simple-minded fox simply watching.

4.8. Red fox

Rather, it seems that this animal is indeed learning, and learning something that it deems important (or at least entertaining). Perhaps this is its first chance in life to closely observe humans. After all, we are weird animals by most measures, with our upright posture and our big heads. We are indeed unique throughout the animal kingdom, and I imagine some of the fox's questions were: Are these cohabitants of the night friend or foe? How fast and agile are they?

What do they smell like? And why are they releasing those bats rather than eating them?

Among the canids of the region, foxes are the smallest in body size (weighing 15 to 20 pounds), and thus occupy a narrow predatory niche. The big brothers to foxes are coyotes (25 to 35 pounds) and their big, big brothers are wolves (80 to 150 pounds), known to outcompete and, therefore, control coyote populations. The consequent reduction of coyote numbers in Yellowstone National Park after timber wolf populations grew after their reintroduction tells this tale nicely. For foxes to survive as the smallest predator in this literal dog-eat-dog world may require a greater knowledge of the environment and potential threats. The expression *cunning as a fox* is well earned, and I wish our observant and educated friend well.

Haunted Nights in Wyoming

Classified as one of the most haunted spots in North America, Fort Laramie National Historical Site looms on the great plains of eastern Wyoming among rolling hills still rutted by the wagon tracks laid down during the days of manifest destiny, when pioneers traveled west to claim a new life. Fort Laramie stood along the Oregon Trail as an outpost, a lifeline to those making a dangerous journey. The spirits of those who passed have left an indelible mark on the landscape, and the energy of this place is thick with human and natural history. The remaining historical buildings at Fort Laramie are much the same as they were in the 1800s. Rangers patrol the grounds 24/7 to keep out thieves and vandals and protect priceless artifacts. Located about five miles outside the town of Fort Laramie, which boasts a population of 350 good people (plus 6 sore heads, as the town sign reads), the historical site is well isolated from any large metropolis. And, according to the rangers, strange occurrences are the status quo of their nightly patrols. Common nocturnal occurrences include buildings that mysteriously become unlocked after being secured at dusk, lights seen in building

4.9. Welcome sign for the town of Fort Laramie, Wyoming

windows even though there is no electricity in them, the clear sound of someone walking across floorboards in a vacant facility—the typical fare expected of haunted sites. Of course, being a scientist, I shrugged off the stories, simply laughing and responding with appropriate curiosity. More or less, I figured these accounts were mostly schemed by bored rangers trying scare tactics on the nutty bat biologist from Colorado. I found out some time later that Fort Laramie is highly considered as one of the most haunted sites in North America in several publications and today on several websites.

I wander the woods at night at Fort Laramie for four years studying the foraging ecology of juvenile and adult little brown bats. During all this time I never see or hear a ghost, but some interesting, even disturbing, incidents happen during my time on the plains of Wyoming. Curiously, a cat—black, of course—shows up on several occasions at my truck parked in the woods along the river. As we are pretty far from the nearest residence, around three miles, I think this is strange, but certainly not supernatural. However, one of the oddest episodes occurs at 3:00 a.m. as I am pulling bats from nets and measuring their wing bones. I hear a noise coming from one of the net stations upriver about one hundred yards away. I go to investigate, and when I arrive at the net I see two piercing eyes peering down at me from a nearby tree. Coming forward, I recognize the figure as a raccoon. At first I'm a bit relieved, but this is not the end of it. You see, I am placing bats in pint-sized ice-cream containers to hold them long enough to collect a fecal sample for dietary analysis. I pile up containers with their inhabitants and then move to the next net downriver to remove bats and do the same. Tonight it is particularly cool (40°F), and the bats in the containers have gone into torpor (a state similar to hibernation, characterized by reduced heart rate and breathing) and are dead asleep. The raccoon, sensing an easy meal, comes along and neatly pops the lid on each container, reaches in, grabs a bat, and eats it. Pleased with its success, it works its way down the line of containers, neatly opening, removing, and eating the contents. All that is left behind for me to find upon my return are pairs of wings neatly parsed out on the ground, the connecting bodies missing. Quite a horrific sight of unexpected carnage is before me, and I feel terrible about my role in the early demise of these bats.

My time spent doing various field projects in the Rocky Mountain West is instrumental to my life as a field biologist and remains today the foundation of career. In more recent years I have had the opportunities to conduct night research with colleagues and my graduate students in exotic areas like the Caribbean, South Africa, Botswana, and China. The uniqueness of these regions are linked for me to the unexpected adventures that come with exploring the world of night biology. And with these thoughts, I bid you "Good night."

FIVE

Nights on the Equator

Ann Kohlhaas

Bzzzzzzz. Bzzzzzzzz.

A common sound in the tropical night. Too common. Much too common. Actually, too common anywhere. At least according to everyone I know, or have ever met, or anyone sane.

Bzzzzz.

Mosquitoes seem to be the bane of everyone who spends time outdoors. Especially in the tropics. Especially at night.

The ecologist in me should defend mosquitoes as part of the greater food web, as essential food for many others. But no. The all-too-human person in me resents the sleep lost, the itchy bumps, and the diseases mosquitoes have already transmitted and will gladly, although unwittingly, transmit in the future.

Bzzzzz.

Much of my research has been in the tropics. With thoughts of my many tropical adventures, I reminisce about things seen and experienced in one of my favorite places on earth.

Bzzzz.

DOI: 10.5876/9781607322702:c05

A Primate's Day

My alarm rings at 5:00 a.m. It's dark outside and it feels a little chilly. At home in the United States, I wouldn't have thought this temperature was chilly, but one gets used to warmer temperatures in Southeast Asia and it feels unpleasantly cool this morning. At this research station on the island of Sulawesi in Indonesia, there is no electricity and it's not worth the time and effort to fire up my kerosene lantern. A couple of candles provide enough light for my few simple tasks. Choosing clothes to wear in the semi-dark isn't difficult. The field clothes I have are all cotton, loose fitting, and characterized by various rips and repairs. They are also all brownish (a hue that is sometimes acquired during the workday). The only choice is whether to wear yesterday's muddy pants or introduce a clean pair to the Sulawesi environment. Yesterday's pants aren't standing on their own yet; they should be fine for another day.

I came to what is now called Bogani Nani Wartabone National Park in Sulawesi to conduct research on the Gorontalo macaque, a monkey species that lives here—and nowhere else in the world—and has not been studied in any great detail. The attraction of this study area and this particular species of monkey was that it was unique and so little understood.

Indonesia is a huge and diverse country, spanning 3,000 miles and including over 16,500 islands. The country literally ranges from the large islands of Sumatra, Java, and Borneo on the Asian continental shelf to the island of Papua New Guinea on the Australian continental shelf. Between those large continental islands are hundreds of miles of ocean and hundreds of islands of varying sizes. Among these oceanic islands is Sulawesi, which was one of the many Indonesian islands where Alfred Russell Wallace traveled, collected exotic specimens, and developed his theory of evolution independently of but contemporaneously with Charles Darwin in the mid-1800s.

After dressing, I go out on the porch to eat breakfast with a lit candle for company. It's still dark, but I prefer to eat outside and look

5.1. Toraut Research Station, Bogani Nani Wartabone National Park, Sulawesi

over the nearby fields as the sun rises rather than stay in my mostly featureless room. The field station is pretty nice. It was built with World Bank money in the mid-1980s, apparently in anticipation of Project Wallacea and its many researchers. Bad timing resulted in the field station being built after Project Wallacea had ended and all the researchers had left. I guess the national park authority expected a continued research presence, but I'm the first long-term researcher to stay here. The few rangers stay in a couple of the support houses. I'm occupying one room in one of the five guesthouses. Occasionally, another room is occupied by a visiting American or Indonesian research associate or, rarely, an intrepid traveler. But most of the time my trusty ranger assistants, Max and Junaid, and I are the entire research team.

Soon after my arrival, Max Welly Lela was assigned to work with me by the park supervisor because he had previous research experience and speaks English relatively well. I later chose Junaid to be an assistant because he impressed me with his work ethic when he was cutting trails for the study area. Both men are fit and strong, very dependable, and pleasant to work with. They are also smart and understand the need to record data accurately, without extra interpretation, embellishment, or attempts to please the "boss."

Max, who has worked for the park for several years, is native to the Minahasan part of the northern peninsula and a Christian. Junaid, who goes by only one name (which is common in Indonesia) is native to central Sulawesi and a devout Muslim.

By sheer luck and circumstance, Clara is my cook. I've never experienced a bad cook in Indonesia, but I swear Clara is the best of them all. I've never told her what I wanted to eat or what to buy at the market—but everything she does is just right. Clara is also an excellent baker and usually includes some freshly baked bread or cake with my meals. Every evening, she brings my cooked food in well-sealed containers along with thermoses filled with boiled water for drinking. Because I don't have funds for electricity or a refrigerator, all my food is kept at room temperature. At home, I would never eat cooked food that had been left out overnight. Here, I eat it every morning and never get food poisoning.

In Indonesia, every meal has rice as its basis. In fact, government employees are allotted twenty-two pounds of rice each month for each member of their family (a spouse and up to two children). That's two-thirds of a pound of rice per person per day! Along with the rice, a typical meal consists of some sort of vegetable and either meat or fish cooked in coconut oil. Coconuts are commonly grown in North Sulawesi, and their oil is plentiful and widely used. Pork is the only food item never served to me because most of the population, including Clara, is Muslim.

After breakfast, I pack my lunch, water, a notebook, waterproof pens, tape recorder, and poncho in my daypack. Binoculars hang from my neck. Junaid arrives promptly at 6:00 a.m. Max has the day off. As the sun peeks over the horizon, we walk down the trail behind my house to the Toraut River. A large bamboo raft is tied to a rope suspended across the river. When I first began my study, we had to wade across the river in a shallow rocky section downstream. It was always an iffy crossing for me. Once, after a heavy rain, Max and I had a rather frightening crossing in rushing chest-deep water, which I hope never to repeat. Now we have a raft, and we cross in the quiet part of the river without incident.

A grid-like pattern of trails crisscrosses the study site. On most days, we search for monkeys, observe them for hours, and record details of their behavior. Today, however, we know where the monkeys are, as we left them sleeping in a large tree at dusk yesterday.

It's now a bit past 6:00 a.m., and we're close to that tree. An adult male makes two loud calls from about one hundred meters north of the tree. The monkeys are already on the move, foraging for the day. We keep a respectable distance when they're active, so they feel safe and do not scatter in a panic. We walk slowly, staying in the relatively open area of the trail. The monkeys can view us and do not perceive us as predators. But although they've seen us many times, they prefer that we keep our distance. We are able to observe their movement in the trees, but we follow them even more easily by their sounds, especially the occasional loud calls of the adult males.

The forest here is "gappy." Although there are many large trees, the canopy is not closed. The many small gaps result from trees falling, bringing down a few adjacent trees in a small area. Many of the gaps are about twenty-five yards across. Some are larger and some smaller. Some gaps are the result of disturbance by humans, but many are natural. Nearly every day, I hear at least one tree fall in the forest. Sometimes a tree falls onto a trail and we have to reroute around it.

The ecological uniqueness of Sulawesi stems from its long isolation, which in turn is the result of the very deep water that has always separated it from other lands. The island's peninsulas are reportedly from at least two chunks of Gondwanaland—the Paleolithic continent that eventually broke into Antarctica, South America, Africa, India, and Australia. Virtually all the life on the island came from ancestors that swam, floated, or flew there. For most land-dwelling organisms, the crossing would have been very difficult. But for those that were successful, their descendants could radiate into new areas and new niches and effectively "evolve in isolation."

The result of evolution in isolation is a large number of endemic species on Sulawesi (endemic species occur in one place but nowhere else). At least 62 percent of the mammal species on Sulawesi are endemic, including two pig species: the rather typical-looking Sulawesi pig and the very odd-looking babirusa, which is nearly bald and has upper tusks growing out the top of its snout and curving backward. There are also two endemic species of dwarf buffaloes, both with very sharp horns that point straight back. In addition, two endemic species of marsupials, called cuscuses, live in trees.

The primates of Sulawesi are all endemic. There are no apes, no gibbons, and no leaf-eating monkeys, but multiple species of tarsiers and macaques are all endemic to the island. Tarsiers are small nocturnal insectivorous primates that occur only in Southeast Asia, including Borneo and the Philippines as well as Sulawesi. Macaques are a widespread group of several monkey species that occurs throughout nearly all of Asia, with one species occurring in North Africa.

The ancestor of the Sulawesi macaques apparently somehow rafted over from Borneo. Sometime after landing in central Sulawesi, the macaque produced descendants, which thrived and flourished as they spread into the four peninsulas and adjacent island chunks. As they occupied these long, relatively thin peninsulas, there were further opportunities for isolation and thus evolution in isolation. Today we recognize seven distinct species of Sulawesi macaques. Not surprisingly, their range boundaries coincide with high mountain ranges, land edges, and low-lying areas.

It's 9:00 a.m. now, and most of the monkeys have settled down. An adult female is grooming an adult male while another female, nearby, occasionally grooms her juvenile, who is lying on the branch. The Sulawesi macaques are all very darkly colored, either dark brown or

5.2. Sulawesi tarsier

black. Some have areas of light brown or gray on their face or limbs. The body of the Gorontalo macaque is a very dark chocolate brown; the hair on the crown of its head is black and longer than its body hair. If I have just the right view, I can also sometimes see a narrow black stripe on its back.

Sulawesi macaques all have rather prognathous faces, which means the face sticks out rather than being flat. They also differ from other macaques in having almost no tail. This is an unusual feature for a monkey, as one of the most obvious distinguishing characteristics between monkeys and apes is a tail. Thus, an early name commonly applied to these monkeys was Celebes apes, but they are certainly not apes.

The female is grooming the male's back as he sits facing away from her. They are on a very large open branch, easy to view, and look very relaxed. The other female is grooming the back of the juvenile, but he is lying on his stomach. After a while, he sits up and grooms the female.

Occasionally, other macaques walk in and out of view. One checks under a leaf for insects and then looks under a piece of bark. For the most part, the area is quiet.

After a couple of hours, the monkeys become more active. The group moves a few hundred yards to a new location. We see many looking for insects along the way. A few find some fruit to eat. Then they settle down for midday resting and grooming. Every fifteen minutes, we record where the monkeys are, their age and gender, how many are visible, their behavior and with whom they interact, what they eat, and the part of the canopy they are using. But much happens between the fifteen-minute data points, so we also make note of behaviors during those times, which include play, agonistic interaction, copulation, and so on.

The clouds have been gathering all day, and it's been completely overcast for a couple of hours now. At 2:00 in the afternoon the rain starts: a few drops at first, rapidly becoming many heavy drops. I put on my plastic poncho and step under a big round palm leaf. The palm leaf protects me from the rain, which otherwise would pelt my head directly. It's a heavy rain, but not as heavy as it gets. Sometimes in a heavy rain you can't see anything at all, but I can still see the two monkeys I have been watching. They're no longer in a grooming bout. They're huddling, not moving at all. They have moved closer to the main trunk of the tree and seem fairly well protected from the rain.

So it goes. It continues to rain. The monkeys huddle, unmoving. I stand under my palm leaf, watching monkeys that do not move. With the protection of a large palm leaf and a plastic poncho, I should stay dry. Not so. The combination of sweat, persistent rain, a palm leaf that is not large enough to shelter me, and a cheap torn poncho ensures that I'll be thoroughly soaked by the time the rain ends. Luckily, the poncho helps to retain body warmth during the rain.

I'm glad this isn't one of those gully-washer rains through which you can't see anything at all; the river swells, and trees will surely fall. Once, during a torrential rain, I heard what sounded like a massive tree falling in complete surround sound. It seemed right above me. I was certain I was about to be crushed. I wanted to live, so I jumped out from beneath my palm leaf and looked up in the hope that I would be able to see the falling tree and run out of harm's way. In my *Twilight Zone* moment, I could hear the tree falling and expected it to crush me. Then, in an instant, the only sound I heard was rain. I did not see a falling tree. There was nothing to do but get back under my palm leaf and wait for the rain to stop. Thirty minutes later, the rain ended. I looked around and saw the fallen tree—only about twenty feet away.

I am once again waiting for the rain to stop, which it does after about an hour. Simultaneously, the monkeys start foraging again. Time's a-wasting. They need to eat as much fruit and as many insects and young leaves as they can before it gets dark, and they need to find a large tree for a safe sleeping spot.

By 5:30, the monkeys have again settled. Many have already ascended a tree and can be seen resting or grooming. By 6:00, it's time for our nightly retreat.

Things That Fly at Night

First Sighting

Sunset is rapidly approaching. It's been a very long day. I can go home and relax. I can also clean up and not be sweat-drenched for the rest of the night. I feel the tiredness that comes of a long day, the giddiness of a good one, and the pleasure of that day being done. Exit the forest onto the raft and across the river. Another successful crossing. I'm still only wet with sweat.

The grasses around the field station glow in the setting sun. With its whitewashed walls, the station resembles a small tropical colony. I look to the sky to finally enjoy its full expanse . . . and there they are.

They're back. Hundreds—no, thousands of them. Filling the sky as far as one can see. When they arrive, they always fill the sky. Huge wings slowly flap and flap and flap. *Bats!* Huge fruit bats! Tonight will be the first of many that they fly this route between a day roost and a supply of fruit. Their dark leathery wings stretch over a meter from tip to tip. I cannot discern any other physical details because they are so high, but I know that the fur on their bodies is a thick fuzzy rust brown. Their large dark eyes look nothing like those of their small bat cousins. In fact, between dog-like snouts and simple standup ears, their faces are reminiscent of fox faces, and they are commonly called flying foxes.

Their flight continues for many minutes. Every time they appear, I think of the flying monkeys in the *Wizard of Oz*—those evil blue flying monkeys. But these are flying foxes, not flying monkeys, and they're just hungry. They keep flying by, too numerous to count . . . until it is dark and they are gone. They've all gone in the same direction—to an abundant supply of fruit somewhere in the distance. It's always a surprise when I see them the first time. They fly over in the evenings for a few weeks. Then they're gone.

THE RETURN

Screech! It's still dark and they're back. If you spend a night near a fruit bat roost, you have no need of an alarm clock. Fruit bats all seem to arrive at their roost at the same time—before the sun rises. And they all seem to be screeching, "I'm home!"

After feasting all night, a fruit bat deserves a leisurely day of rest hanging upside down in the sun for extra warmth or in the shade to cool down. Well, forget that. No resting for fruit bats. Day roosts are apparently not for resting. No. Day roosts are for fussing. Fussing about everything! The bats constantly move and change positions. Sometimes two of them have tiffs, and sometimes an entire group seems on the outs. Over and over again. All day long. Very few are ever observed to be simply hanging.

And the noise! Every action deserves a screech . . . a separate one by every bat involved. *Screech!* A walk through a day roost of thousands of bats is an experience of surround-sound screeching. It's a cacophony that could certainly induce madness. Like a rock band that spends more on speakers than music lessons.

Things That Slither

Not surprisingly, there are snakes on Sulawesi. When I first arrived and told people that I was going to study monkeys in the forest, several advised me always to wear a hat in the forest. Why? Because of snakes in the trees. When I asked how a hat would protect me against tree-dwelling snakes, I was given two different explanations. One was that if a snake tried to bite me from above, it would effectively bite the hat and I would get away. The other explanation was more common. If a snake fell or jumped out of the tree onto me, it would hit the hat and drop away without biting me. While both scenarios are possible, they certainly seemed unlikely, and I declined to wear a hat in the forest.

There are definitely snakes in the forests of Sulawesi. I don't go looking for them, but I generally see at least two per week. Since the research station is on the edge of the forest, they also visit us there. One day, I notice a shed snakeskin in the window slats of my room. I wonder when the snake was there. I wonder harder where it has gone.

One snake I watch for is the giant python because I know that the longest snake, according to the *Guinness Book of World Records*, was a python from Sulawesi—caught in 1912 and about ten meters long. I have asked locals if they had ever seen any big snakes. Most say they knew about a man killed by a large python. Not too surprisingly, it isn't a story anyone has direct knowledge of. There was a man, a farmer (who isn't?), who was missing. A large bulging python was found in the area and killed, and the farmer was found inside the snake. No one can name the exact town, but it was "over there."

As much time as I spend in the forest, I catch a glimpse of a small python only once. However, a large python does come to visit us at the research station one evening. We are having a small party at my house, drinking saguer, a fermented drink made from palm sugar. No one hears or sees the python that evening. The first hint of it the next morning is an incessantly barking dog. Finally, someone investigates why the dog is barking and discovers the python—a large specimen, but not record breaking. It's about four to five meters long and as big around as the palm trunks that support the roofs of our houses. Having killed one of Clara's goats, it is in the process of trying to consume it. Its luck runs out that morning. Several Christian men (Muslims don't eat snake) doing construction at the station kill and eat the python. Clara cooks the goat.

Not in the House

Time to get up. I'm feeling warm, but not too bad. I've got a fever. The real worry is that the fever is alternating with chills. I'm concerned that I may have malaria. I've been back in the field for over a month after some time in the States and have taken my chloroquine as prescribed. I'm worried because there's supposed to be a chloroquine-resistant malaria in this area. Chloroquine protected me last year, but I may have run out of luck this time. At least I don't feel too bad right now. It's just a low-grade fever. I had let my park ranger assistant know that today would be a day off. I hoped that time off would help me shake whatever the bug was. Please don't be malaria!

I'm finally getting up. Feverish but still okay. Time for some breakfast and rest. On with some clothes. On with the ubiquitous rubber sandals. Time to go outside and greet the sun.

Slap! What was that? It sounds like something has fallen on the floor in my room. I quickly walk back inside and look down the length of the room and there it is. A snake on the floor by my bed! It must have fallen, or more likely jumped, from the slats. I am surprised to see the snake, but considering that there are no screens and the walls of the room are half-solid and half-open wooden slats, I probably should not be.

A snake in my room. A snake about three feet long. Not something I want in my room. Even if most of the snakes here are harmless, it is an

5.3. Snake that invaded my cabin

uninvited and unwelcome guest. It's got to go. First I need something to shoo it out. The broom! Where's the broom? On the porch! Got the broom! Now to get that snake out of my room.

Where is it? Not by the bed anymore.

Is it under the desk? No.

Is it under the other bed? No.

Is it under or behind the cabinet? No.

Is it hiding behind my luggage? No.

Where is it? Did it go back out through one of the slats, which start a few feet up the wall? A difficult thing but not impossible. Unlikely, though.

I am feverish. I'm not seeing things, am I? There was a snake. I'm sure there was a snake. I'm not that sick. But where is it? Okay, it's not by or under either bed, or the desk, or the cabinet, or the luggage. So where could it be?

It wouldn't have gone into my bed. Would it? Surely not. Well, I hope not. It was last seen by my bed. I should check. Just to be sure. I approach the bed slowly. No sign of a snake there. It doesn't seem to be in the sheet or flannel blanket, or in the mosquito netting. It wouldn't be under the mattress, would it? That wouldn't be comfortable, between a foam pad and wooden slats. No. But I should check anyway. Yes, I'll check for my own peace of mind.

I gingerly grab one corner of the thick foam pad, still brandishing the broom in the other hand. For a split second, I see the snake lying on the slats. It had been between the wooden slats and the foam pad. Now it is already several meters away. I chase after it with my trusty broom. In almost no time at all, the snake is out the open door . . . and under the closed door of the next room. Great! Now there's a snake of unknown species and venom capacity in the adjacent room. And the door is locked. It didn't look like a cobra, which are nearly black here, but I'm still not certain that it isn't venomous and I don't want a snake in the house anyway!

One of the park rangers walks by, and I tell him about the snake and ask him for a key to the next room. He seems incredulous. Either he doesn't believe there is a snake next door, or my rendition of events was unclear. Both explanations are equally plausible. Within a few minutes, the ranger brings the key and we open the door. We look around carefully and—*zoom!* There's my snake again. I feel exonerated. Yes, I'm not so feverish that I'm imagining snakes. With my trusty broom, I herd the snake out the door, onto the porch, off the porch, and, finally, away.

I was happy that it was not a cobra or another venomous snake. Even so, some lessons take only one event to learn. Monsters might not live under my bed, but snakes could. For the remaining year of my field study, I check for snakes under my bed every night before retiring.

5.4. Wagler's pit viper

SOME SLITHER, SOME FLY

One of the most beautiful snakes that inhabit the area is the Oriental whip snake—about three feet long, very thin, and very green. I would see it sunning on a green bush, very calm and beautiful. Its thin triangular head terminates in a very pointed snout. Most surprising are its horizontal pupils, which reportedly give it some binocular vision—an aid in hunting active prey. It is also a rear-fanged snake and mildly venomous to help it kill lizards, its usual prey.

Wagler's pit viper is also a very beautiful bright green. Its coloration matches the vegetation well, making it difficult to spot. I wonder how many I have missed. It is very disturbing to see one at the edge of the trail, looking like it could strike at any moment. I usually see it in the forest on the ground or a couple of feet off the ground. Once one was in a bush near my house. Its wide triangular head and the large obvious pits in front of its eyes for sensing warm-blooded prey are a clear indication that this is a very dangerous front-fanged pit viper. It is only about two feet long and heavy bodied.

I nearly step over a cobra in the forest one day. Just before doing so, my brain registers that the "vine" wrapped on the fallen tree that I'm about to step over is a snake. I jump back and nearly knock over Max, who's following me. We watch the snake disappear beneath the tree. We try to coax it out to see it more closely, but without success. Just as we are leaving, Max hits the fallen palm leaf that we have been standing next to and out slithers the snake—a five-foot cobra that gives us a brief hood spread in response to our jumping out of its path before quickly slithering away.

Flying snakes are also present in Sulawesi. They don't technically fly; they flatten out and glide from one tree to another. Their climbing ability is phenomenal, as they can stretch out about half their body to the next branch and use just a small portion of the branch to similarly reach up to the next one. They climb trees to hunt for lizards, which they kill with the mild venom that flows down their rear fangs.

Nighttime Disturbances

A typical evening is rather uneventful. I bathe at the river by scooping up water in a small pail and pouring it over myself. It's a pleasant experience on a hot day, but much less pleasant when it is cool. It's nice to put on a clean outfit after sweating in my field clothes all day. Clara has brought my dinner and breakfast and lunch for the next day and taken the previous day's containers. I light the kerosene lantern and sit on the porch to eat my dinner. A short-wave radio provides news from the BBC and Voice of America. It's been a long day and after making some notes, I put out the lantern and head to my bed.

Before falling asleep, I read for a while by candlelight. It's not the best setup, but it's relaxing and the candle gives just enough light through the mosquito net. The book isn't that important. It's whatever novel I happened to pick up in Singapore or Jakarta or one left by a visiting associate. As I grow sleepy, I blow out the candle and make sure my mosquito net is properly tucked in. As always, the mosquitoes are here. *Bzzzz.*

RATS

What's that noise? Oh, no, not again. It's amazing how loud rats can be in the middle of the night. It might be partly due to the acoustics of this large, relatively bare room. You would think silence would be the rodents' key to success in a human habitation, but they don't seem to try to be quiet when they are in my room. Their strategy seems to be to run quickly (and noisily) and hope to find something good to eat.

Foraging in my room is not productive for rodents. All my food and water are locked up. Nesting materials are in short supply. Or at least they have been since I found a mouse nest in the paper supply in my desk drawer. The mice aren't so loud, but they can still ruin a night's sleep.

Fortunately, only about one rat enters my room every month. I learn quickly that rats must be caught right away or the noise continues, seemingly endlessly. After a long day in the forest, I need a decent night's sleep. So I have effectively armed my room with two traps and usually catch the rat within a short time after it enters. After the rat is trapped, I have to get up to deal with either a dead rat or a soon-to-be-dead rat. I have to kill it quickly if it is not already dead, and then dispose of it at least one hundred yards from the house.

There are numerous types of rodents on Sulawesi, including many rat species related to the common household rat, the roof rat, found in the United States and Europe. Roof rats are quite fond of living with humans. Sulawesi has the same species, but also several others in the same genus that are very closely related to the roof rat. All the rats I have caught look basically like the roof rat, but I can't be absolutely sure, as I don't know the distinguishing characteristics of the other related species in the area.

One of the early captures occurs the night before I have a day off. I keep the dead rat, deciding make a study skin of it in case it's an unusual species. I'm out of practice making study skins and don't have the proper tools or equipment, but I make do. The basic technique is to separate all of the body except the foot bones from the skin, and then re-create the general body shape within the skin with a wad of cotton. The feet and tail are stiffened with wire.

It takes me a couple of hours to make something like a study skin. Then I pin it carefully on a piece of cardboard so it will dry in the proper shape. It's important that the skin dry out and also that it be protected from insects. Even in a tropical rain forest, there are some dry afternoons, and that afternoon is one. But the humidity is high, and the study skin needs a few more days in a relatively dry area. Also, insects are abundant, and I have to keep the skin where they are least likely to eat it. That evening, I put it on the top of the cabinet in my room. Although that spot is never sunny, at least it's away from many insect sources and less moist than anyplace outside the room at night.

Within a day, ants have found my rat and eaten its ears and some of its feet. So I take it outside and put it on the ledge that stretches around the house. The specimen is damaged at this point, but I don't want ants in my house and storage cabinet. The rat stayed on the ledge for several days, earless and unmoving and slowly drying—a very pathetic study skin. Then one night it disappears. The cardboard remains. The pins are mostly in place. But the rat is gone.

I wonder where my rat "ran off" to. It's hard to imagine that any human wanted it. I like to think that some predator took it away, happy to find such an easy-to-catch prize—then very surprised to find its insides were cotton.

Rumblings

It's the middle of the night . . . and suddenly my entire bed is shaking—the entire house is shaking. Earthquake! I've spent more than a year here and this is the first violent earthquake I've experienced. The others were very mild earth rumblings. This one is scary. I quickly jump out of bed, disentangling myself from the mosquito net, grab and put on some proper clothing from the clothesline, and start for the door, effectively kicking my rubber sandals to the door rather than putting them on.

I make it out to the concrete walk just in front of the house as the shaking stops. Now what? Is it going to shake some more or can I go back in? The house is okay. All the houses at the station look okay. It's the middle of the night and dark outside. Maybe I should sleep on the concrete walk. No, it's cold and wet. The porch isn't wet, but it is still in the danger zone. Maybe I can drag my bed out here to the sidewalk. No. I'm not that strong. I don't see anyone else; the rangers live on the other side of the station. After about ten minutes of indecision, tiredness and a bit of bravado lead me back into my house and to bed.

Living along the Pacific Ring of Fire, one has to expect a little earth shaking. The Pacific Ring of Fire is composed of geologically active zones encircling the Pacific Ocean. In these areas, the plates of the earth's crust grind against one another and cause earthquakes and volcanic eruptions. I'm a little surprised that so much time had passed before I felt a real earth jolt.

Oh, no! The house is moving again. It's still dark, and I'm awake again and moving out the door even more quickly than before. I guess I should have stayed outside last time, but it was nice to get a couple more hours of sleep. Same thing again. I'm outside. It's quiet and dark. No more shaking. No people. Just me and darkness. At least dawn is coming soon. Surely this won't happen again tonight. I'll go back in and sleep some more.

The next morning, we go to work in the forest as usual, except our conversation in dominated by last night's earthquake. The rangers have experienced quakes before, but this was a rather strong one. The ranger house on the crest of a small hill really shook, they say. A couple of the rangers slept outside the rest of the night or didn't sleep at all.

When we get back from work and look about the station more closely, some damage is apparent. Although everything is still standing and functional, a couple of rooflines are no longer straight. My house looks basically okay, except that the palm log supports for the roof have moved a few inches on their bases.

In the following days, we occasionally feel some latent earth rumblings. Those nights, I sleep fully clothed—ready for a quick exit.

Shakings of Another Sort

The morning I find the snake in the house is just the beginning of my feverish episode. As that day progresses into night and into day again, my health steadily declines. I develop the cyclic fever and chills that typify malaria. The first day or two, the fever is low, and I'm hopeful that I am ill with something else, which will just go away. Then one night I awake, shivering with violent and uncontrollable chills. I have never experienced shivering of this magnitude. I am bathed in sweat. My T-shirt is soaked in a strangely sweet-smelling sweat, exacerbating the unbelievable cold I feel. I change into a dry T-shirt. I also take some aspirin in the hopes that it will help the situation. After about thirty minutes, the chills subside. It's a strange sensation of relief to go from uncontrollable and dramatic shaking to feeling just vaguely ill.

Unfortunately, over the succeeding hours, the feeling of mild illness and fever steadily develops into a much worse fever, eventually culminating in another soaking of sweet sweat followed by bizarre shaking and chills. The symptoms seem to indicate malaria. I search my brain for other possible maladies, but the cyclical nature of fever and chills match malaria's usual symptoms. I'm not sure which strain of malaria I have, as my fever-chill cycle is rather rapid—under twenty-four hours.

Beside the obvious, I have another reason for hoping that I don't have malaria. I don't want to take Fansidar—one of the drugs available to treat malaria—because there are reports of people developing problems after long-term use. The other medications in use in the area are chloroquine (which I am already taking) and doxycycline (an antibiotic, which has to be taken daily). Daily use of an antibiotic over many months would likely have given me other problems.

I endure the fever and chills for a couple of days, hoping they will stop. They don't, and I'm feeling much weaker a few days later. I can eat during the intermittent lulls but not much, and I become very picky about what I can stomach (nothing greasy). I have a sore back, but that may be because I have been sitting and lying around on basically uncomfortable furniture. There are no doctors near the park, and options for going into town are limited. I also don't really relish the thought of several hours in a bumpy crowded truck.

Finally, I decide to try Fansidar. I worry that in another day the illness will progress and I will not be able to make a sound decision. I'm feverish but, I reason, I have never had a bad drug reaction, this is a one-time dose, and my odds are better with the drug than another night without it. So of my three pills of Fansidar, I take just one, thinking that I'll see how I react. Then I realize how stupid that is. I need the curative dose, not a little knock-back dose. I take the other two pills.

I heard a few years later that Fansidar had been taken off the list of antimalarial drugs commonly prescribed by U.S. doctors. Apparently, the numerous allergic reactions and the tragic chronic reactions had taken their toll. But for me that day, Fansidar is my hero. By the next day there are no more dramatic fever or chills. I have a low-grade fever and a feeling of weakness and mild illness for two or three more days. Then one morning I awake with a tremendous feeling. The fever has broken. It is totally gone, and I am feeling absolutely fine. I know that the illness is over.

Good Night

Bzzzz.

Sulawesi seems long ago and far away now, though it is never far from my thoughts. Even here in California, the mosquitoes *bzzzz*, but these carry the potential to infect me with West Nile virus rather than malaria. Snakes still surprise me in the field, but they don't visit my home—I no longer check for them under my mattress. The only bats I see in the evening are small ones that hunt insects (mosquitoes, I hope). But the monkeys aren't here. They're high in a tree sleeping in Sulawesi.

SIX

Do Not Go Gentle into That Tropical Night

LEE DYER

I have always been afraid of the dark. So not only is it ironic that I work at night a great deal, it is also odd that I work in the dim dank understory of tropical forests. Despite the numerous gaps, edges, rivers, and clearings, it always seems dark in the rainforest, where the green walls press in on all sides. Nights in those green cages offer a special brand of lightless energy—something more than a simple word like *dark* can manage to capture. It is true that the small details of life that I like to study are best seen in direct sunlight, with good fiber optics or with a brilliant spotlight, but insects and other small movers and shakers of biotic communities don't always perform well under such circumstances. As a consequence of working in tropical forests, I have endured many night excursions and many hours of sitting in shadowy places, wondering what was crawling up my leg or tickling my neck. I've had to suck it up and deal with my fears.

I think about all this while I walk through known territory one night—a fragment of Costa Rican rainforest, La Selva Biological

DOI:10.5876/9781607322702:c06

6.1. Costa Rican rainforest

Station, where I have worked for years. Another invertebrate ecologist, Eileen Hebets, has lured me out to find amblypygids (tailless whip scorpions) and to observe something about them—I can't remember exactly what was behind this night excursion. I do know that walking through a rainforest at night is part of being a tropical biologist, and it is a world like no other. As I try to keep up with her quick pace, I can't help but wonder about all the eyes out there watching me and I can't forget numerous great stories from fellow tropical ecologists who spend nights peering into the dark woods.

My favorite story has to do with an unintended encounter with spiders. Tom Walla is a pretty tough naturalist, with years of experience working in various ecosystems in Ecuador. On one of his many excursions, Tom and a group of biologists, all under the influence of some lowland forest plant brew, were restlessly wandering around the forest at midnight, examining animal eyeshine, luminescent fungi, aerial caterpillar silk, and deep holes in trees. For some unexplained reason, they were also stripped down to boots and hats. Some biologists like to be nude and intoxicated in the forest at night, I suppose. After some time, they realized they were completely lost and that it was time to extract themselves from the situation. Tom, who likes to be in charge, was near the end of the line of lost souls, wishing that he were leading this self-rescue attempt. His constant string of loud complaints did not go unnoticed, and soon the leader of the group, Harold Greeney, was considering giving in and turning over the navigation to Dr. Walla. Just then, Greeney came across a massive nest of social spiders—thousands of little spider eyes peered back at his flashlight. Arachnophobia is a very common phenomenon, even among entomologists, but social spiders strike a special fear into even the most undaunted hearts. And encountering them at night is, shall we say, special. Greeney stepped up to the very edge of the spider nest, then stepped aside. "Okay, Walla, you're in charge. Come on up here and lead the way." I've often seen Tom take charge of situations, and I can imagine his immediate run to the front of the line—and straight into the middle of the huge social spider web. However, I cannot imagine being covered by the obscene number of spiders that besieged Tom's naked body that night. His writhing dance in the mud and his eerie screams have been described in detail by several individuals, and all descriptions are corroborative—and all make my skin crawl. Now, as I lumber through the dark in Eileen's wake, I shudder at the thought of the social spiders, stopping to check for critters on my body. When I catch up to my companion again and tell her the Walla story, she isn't amused; in fact, she seems a little annoyed. I think she has no patience for arachnophobia and perhaps has a secret hankering to roll around in the mud naked covered with amblypygids.

There is a short silence, during which Eileen must be pondering the tale, for she asks after a while, "Why they were naked? I mean, was it some sort of macho romanticism combined with stealing some native religious ceremony? It sounds pretty dumb."

I shrug. "I think they were just having fun in the rainforest at night." Suddenly I notice a big clump of red and yellow caterpillars eating a cycad. "Hey, check out these hairstreak caterpillars— they are super-cool. You know, they sequester this compound called cycasin, which is pretty toxic. We should see if the amblypygids will eat them."

Eileen is genuinely interested in the group of caterpillars, but she does not agree to try the feeding experiment. She is focused on her research objectives and does not have many nights left to complete her work. We move on toward our objective, and as we walk I notice a pair of large black ants crawling up a thick vine to the invisible canopy. They are graceful creatures, well built, with shiny black bodies, some of them wearing yellow boots, some with wings, and some with stingers perpetually ready. The giant tropical ant, *Paraponera clavata*, is feared by most people who encounter it. The many common names for this ant reflect the very painful sting it inflicts. The English versions of these names include bullet ant, twenty-four-hour ant (the time frame of the pain), forty-eight-hour ant (how long you should stay drunk after being bitten by it), bad woman, and others that may be used in other parts of its range or by surprised visitors to the rainforest. Despite their painful sting, these giant ants are my favorite animal. *Paraponera clavata* is supposedly active mostly at night, but in my research, I have found that it is equally happy to look for food during the day, especially cloudy days. In fact, I have spent hundreds of hours observing the foraging patterns of these ants, recording their predatory habits, food preferences, and activities at all hours of the day and night. One of my conclusions from years of research is that bullet ants remove a significant proportion of herbivorous insects (especially leaf-cutting ants) from rainforests, and consequently it is partially because of them that these forests are so green.

I remember the first time Christine, one of my field assistants, visited the rainforest in Costa Rica and encountered *Paraponera clavata*. We arrived at the field station at night—just twenty-four hours into her first trip outside the United States. I was eager to start collecting data on the foraging behavior of the ants. On the bus ride from the airport to our station, she had read much of the "*Paraponera* book," which describes the impressive size of the ant, the incredible pain inflicted by the sting, and some of the most extreme reactions (including death) that some have experienced as a result of one or multiple stings. I foolishly left Christine by a tree to collect data on these ants, and when I returned more than an hour later, she had not moved a muscle. She was frozen in place, watching the ants, hoping that they wouldn't notice her, mobilize their forces, sting her thousands of times, and carry her down as a dinner item for their underground maggots. I don't think she has ever forgiven me for that night in the field.

My first glimpse of these ants came in daylight, and I recalled a friend's description of the ants quickly recruiting nest-mates as soon as they encountered a person. Best to flee immediately. Of course, I found that it was not like that at all. It's true ants recruit others to help them harvest food sources and will gather in big groups in order to tear up caterpillars or other arthropods quickly, but they do not assemble in large groups to attack humans. The lack of interest in humans as a food

6.2. Bullet ant, Costa Rica

source is fortunate because a single sting from this ant is all it takes to ruin your day.

My daydreaming about bullet ants and terrified volunteers is interrupted by a bizarre loud yell from a bird that sounds a little like a poor-will. The paraqui. Or is it the mythical goat-sucking monster from Mexico, the chupacabra? The paraqui's call usually elicits a fond response from biologists, but parents invoke the bird to strike terror into the hearts of children who have misbehaved: *The chupacabra will come get you tonight if you don't behave. Do you hear him out there?* But the chupacabra really does not compare to the nighttime terrors of the arthropod world unless you're a goat. To me, the paraqui's call is quite comforting, as is the visceral growl of another night bird, the potoo, which greets me next. I suppose the potoo's odd noises have scared a good number of unsuspecting forest visitors. It sounds a little like a mad dog coming to get you from a direction you can't determine.

The only time that a similar noise has scared me was one late night in Quito. I was carefully making my way down a shadowy street to the hostel where I was staying when I heard what I thought was a potoo, or was it an insect? Investigating, I thought it might be coming from a drunk, crumpled on the ground. I glanced at his body, noticed a few insects crawling over him, and then heard him make the low guttural noise. I quickly left the area. The previous year in Quito, I had been robbed at gunpoint, pistol-whipped, and terrorized for several hours by a group of thugs in my hotel lobby at 2:00 a.m. The attackers made me and others lie on the dirty floor, where I watched a pool of blood accumulate from a nasty cut on my head as cockroaches scurried around us. I lost all my possessions that night and was injured as well, but I felt very fortunate to live through the incident. However, the assault put another spin on night fears, and I am usually a bit jumpy in dark alleys late at night in Quito. Frankly, I'd rather take my chances at night in the forest with the ants, spiders, and snakes.

The calls of the paraqui and potoo are now far behind. Eileen and I have stepped off the trail and are making our way to one of her marked trees. She suddenly yells something incomprehensible and jumps toward a tree, dropping her flashlight and grabbing at something on the trunk with both hands. I find this action far more frightening than the ants, spiders, and birdcalls that we've encountered. What if a big dangerous snake, like a bushmaster, is waiting at the base of the tree? Snakes are one of the real dangers of working with insects at night in a lowland tropical forest. I don't know how many snakes I've stepped on, many of them quite venomous. I go over to examine what Eileen has caught. It is, of course, one of the amblypygids that she is studying. These night creatures, which look like large scorpions without tails, walk on six of their eight legs and sport two long whip-like front legs that act as sensory mechanisms. They use the whips to locate hapless insect prey, which they shred between two massive spike-covered legs in front of their jaws. Amblypygids do not possess poison glands, stingers, or noxious gases, but they do have a strong bite and they are

6.3. Whip scorpion

definitely creepy looking. An amplypygid was used in a Harry Potter movie as the recipient of evil spells in the classroom. I think the moviemakers, looking for a scary-looking "bug" to terrorize audiences, gave the role to a tailless whip scorpion.

Apprehensively, I hold the amblypygid for Eileen while she measures various parts of the beast. This scenario is repeated again and over again, and I soon become immune to all the insect noises, the eyes watching me, the amphibian calls, and all my primal night fears. I do not notice the spider that has crawled inside my pant leg. I ask Eileen questions about her amblypygid study and we both get lost in that glazed academic world of discussion, thought, and exploration. Our discussion of how arthropods would evolve mechanisms for breathing underwater is charting intellectual territory that is much like the murky forest that surrounds us, and the conversation effectively filters out the active night.

What really lured me into the jungle this particular night to handle beasts from Harry Potter movies were Eileen's ideas about how arachnids may be able to breathe underwater using the same mechanism as many aquatic insects—a thin film of air called a plastron. Eileen tells me again of how she first discovered the amblypygid's ability to stay submerged for hours, using a plastron held around the insect's body by skin modifications such as hairs and ridges. She had some in captivity for behavioral observation and was attempting to force them to feed in full view by making a cage with slippery walls (Teflon lined—slippery even for an arthropod) and a single rock surrounded by water. Eventually, the creature is forced to stand on the rock and do its thing. But her first amblypygid immediately hid under the rock and never came up for air. She spent hours watching it until it became obvious that this was no ordinary arachnid; it could breathe underwater. After staying up all night without taking her eyes off her organism (not really a difficult feat for somebody who spends long nights looking for arachnids), she was amazed to discover that it was completely healthy after twenty-eight hours of continuous submersion. Eileen realized that she had discovered something more interesting than the behaviors she had set out to study. She temporarily shifted her research to investigate the special outer covering of these amblypygids that created the so-called plastron and enabled them to breathe underwater for so long. Eileen's discovery was significant because arachnids were not thought to have plastrons, and the revelation that they did has numerous ecological and evolutionary implications. Eileen and I discuss these implications for a while longer and then get back to catching and measuring the nasty predators.

A few hours drag by, and I am tired of the amblypygids. It is 3:30 a.m. and I need some sleep before starting my own fieldwork at 7:00 the next morning. I convince Eileen to finish her work with somebody else the following night, and we stumble back toward the research station.

As we walk, I look for more caterpillars but see only a few inchworms hanging from threads. Years ago, on my first night walk in the rainforest, Bob Marquis, an expert on tropical herbivory, told me that caterpillars are best found at night because that is when many species feed after hiding from predators and parasitoids all day. Since that time, I've collected well over 3,000 species of caterpillars—probably millions of individuals, and I hardly ever collect at night. It is ironic that I collect immature moths almost exclusively in the daylight hours because most lepidopterists collect moths at night—at black lights, mercury vapor lamps, and various other light traps. My method of collecting moths in the daytime, via rearing the caterpillars to adults, is extremely effective, and I often rear new species that are never found at lights. I really do not believe that caterpillar hunting is best done at night, but I occasionally like to try it out, even if I am afraid of the dark.

Eileen heads over to work the rest of the night in the laboratory, and I walk slowly across the large suspension bridge toward my cabin. I stop at my favorite night view, looking down the Rio Puerto Viejo—the heart of darkness—and then I suddenly notice an uncomfortable feeling in my groin. In fact it is more than that—it is a disturbing mixture of jabbing pain and weird movements around my crotch. Once inside the cabin, I immediately disrobe, and to my horror, I see that my penis is bleeding. I've never seen such a thing, and it captivates my attention for a few terrible seconds; then I peer into my underwear. There it is. A big red and hairy spider in all its glory. I have no idea how somebody can walk around with a big spider biting his penis and not realize what is going on, but stranger things have happened at night in the rainforest. I yell something and quickly flick out the spider, which scurries off, unharmed. As it is scurrying, I make a quick mental note, "red ctenid spider, five inches in diameter, somewhat hairy." I admit that this is a harrowing experience, but it has been a long night, and breakfast will be soon, so I sleep fitfully for an hour before heading over to the cafeteria.

The sun has crept up and morning birds and insects have taken over from the previous night's menagerie. I stumble to a table with a plate

6.4. Nocturnal light trap covered in Costa Rican insects

full of rice and beans and sit next to Eileen, who looks pretty chipper this early in the morning, considering she is a night worker.

"A spider bit me," I mumble.

"Really?" Eileen's interest is perked. "What kind?"

"Ctenid."

"Eww, that's bad."

I raise my eyebrows. "Bad? What do you mean?"

"Well, ctenids have pretty bad venom, you know. Sometimes it can be necrotic. Other times it makes you really sick. The reactions can be delayed for a while too. Really, this is interesting because not much is known about ctenid venom. Where did it bite you?"

I eat a few forkfuls of rice and beans, avoiding the question. "What are you doing today? Looking for more amblypygids?"

Eileen looks at me for a minute and then grins. "Oh, no. You could be in big trouble. Can I see?"

I shake my head and respond, "Well, I've heard that ctenid bites give you special powers."

It is time to get to work, so I shovel in the rest of the rice and beans and head toward the suspension bridge. I do notice a little dizziness as I near the center of the bridge, and then suddenly, out of nowhere, a fireball screams across my immediate field of vision. Then an amazing burst of color immerses me. I've seen this show before: hallucinations. Eileen's comments about how the effects of ctenid venom can be delayed echo in my head, and I turn back toward my cabin. Things become black and chaotic. A Dylan Thomas poem shrieks at me, "Rage, rage against the dying of the light . . . "

Hours later, I wake to the sound of some *Leptodactylus pentadactylus* giving their whooping frog classic introduction to another night. The lights are out in the cabin, and I am in bed, hot and sweaty. My head feels a bit light, but everything else appears to be in place and working well. I can hear a group of students off in the distance, looking at moths and beetles at the blacklight sheet. I get up and peer out the window. It looks like another stellar evening, so I go outside and am surprised to see the constellation Orion lying on his side in a relatively clear sky. Stars are a special treat in rainforests, so I sit on my step and gaze up at various constellations as I begin to enjoy a relatively quiet night in the tropics.

Anywhere in this world, when I am looking up at the stars, the moon, or a dim canopy overhead, I reflect on nighttime events of my past and how they have influenced my life. Some of these events flash through my mind, which is not so fuzzy anymore. I remember running, terrified, from lamppost to lamppost as a child, forced to walk home on extremely dark nights. I think about my days hiking the Pacific Crest Trail and realizing that all the desert portions are best walked at night, when the temperature drops to a reasonable level and the nightlife, especially the insects, really starts acting up. I think about several odd brushes I've had with thieves and murderers—always nighttime incidents. And there have been nighttime romances on tropical beaches

6.5. Dark understory in the tropical forest at night

and rock-climbing trips, in treetops, canopy walkways, and regular old bedrooms. The stars smile down on me as I see clouds moving in from the east. I smile back. Is there really a better time than night to enjoy the tropics?

SEVEN

Nights
From South to North, Hot to Cold

JAMES C. HALFPENNY

Rattlesnakes, Beetles, and Packrats

It was darker than a black desert night in Texas. There was a new moon, thick clouds, and my AA-battery ultraviolet light shed a beach ball–sized circle of light. Beyond the light's reach, I heard it: the fast rattling buzz of a black-tailed rattlesnake.

When I started this project, I had purchased the finest of rattlesnake-proof chaps. At this moment, I considered the construction of these chaps, which cover neither your crotch nor your backside. Chaps work great when you are standing, but if you're crawling on hands and knees, as I was doing, both your butt and your nose are fully exposed.

As calmly as possible, seeing as I'm at eye level to a rattlesnake, I asked Scott to pass me our "serious" light.

To explain why I was face-to-face with a rattlesnake in the dead of night, we have to go back in time. Scott Elias and I were working to reconstruct the records of the climate 40,000 years ago. Scott is probably the world's foremost expert on fossil beetle elytra—the hardened

DOI: 10.5876/9781607322702.c07

7.1. Packrat

wings that form the covering of the backs of beetles. Climate 40,000 years ago, beetles, and hardened wings are all part of the story, and here's another: packrats, also called woodrats, defecate and urinate in the same place, creating a pile called a midden. Middens desiccate in the desert heat. Piles of feces deepen over time, and in sheltered caves, the material at the bottom of the pile may be 40,000 years old. Paleoclimatologists burrow into a layer and date it by accelerator radiocarbon analysis. Pollen analysts study the pollen from that layer and, based on the plants that were present, they estimate what the temperatures and precipitation were when the packrat relieved itself. Layer by layer, a climate record can be reconstructed. In some areas these records may go back hundreds of millennia, but where we were the limit is about 40,000 years.

Works fine, except for one hitch. Say there is a juniper or some other long-lived plant in the area, but the climate becomes so harsh that when its seeds sprout to seedlings they die. Junipers are tough—and persistent. For the next couple of hundred years, the juniper puts out pollen but no newly germinated junipers survive. The pollen analyst is led to believe the climate is juniper-conducive when it isn't. Pollen provides only a coarse record of climate change with a resolution of decades, not years.

That's where Scott and his beetles come in. Many beetles are highly "climate particular," residing only where the temperature and precipitation suit them. If the climate changes, the beetle unfolds its wings and leaves. A new beetle species that likes the new climate comes in. Beetles, then, are a very sensitive indicator of climate. By examining the beetle elytra in a midden layer, Scott can tell which beetles inhabited the layer and can estimate the climate at that time.

What do beetles and packrat scat have to do with working at night and meeting up with a rattlesnake? Well, the intriguing question to me was whether packrats could bias the paleoclimate record. Let's say there was a tall hill rising 2,000 feet from the river to the top. Might a packrat search up the hill until it found a pretty beetle (whatever pretty is to a packrat) and bring it back down to its nest, thereby biasing the record? My job: to learn where packrats forage.

Aware of my expertise in animal tracking, Scott figured I could just track them to determine where packrats went. But there's a problem. A packrat's front feet are about half an inch long and its hind feet about one and one-half inches long, and are both mostly covered with hair. Those factors make the tracks difficult to see; and to make my job even more difficult, I had to track them over windblown desert, through yucca, cholla, jumping cholla, and beavertail cactus.

Aware of the inherent difficulty in following a packrat trail, I opted for a practical method. First I livetrapped packrats. Then I dropped each packrat into a plastic sack full of fluorescent powder and shook the sack (the nontechnical term for this method is "shake and bake"). When released, the packrats would shed fluorescent powder for about three days. Using a black light on the first night, I could find perfect footprints. On night 2, I could find a well-defined trail of powder. By night 3, I would be down on my hands and knees squinting under the black light for pinprick-sized specks of powder. Working during the dark of the night, I would place a flag wherever I found powder. In the daylight, the flag locations revealed the travels of my nighttime weather-rat.

That's how I found myself nose-to-nose with the rattlesnake. Rattlesnakes use dark nights to surprise mice and packrats—and the occasional weird scientist. Responding to my request to hand me the "serious" light, Scott passed it to me between my legs. Three feet to my side lay a four-foot rattlesnake. With due respect, I edged backward. I met a lot of black-tailed rattlesnakes in those two field seasons. Thankfully, they're tolerant types who don't get easily riled.

I also learned another interesting fact: scorpions naturally fluoresce. Once when I was crawling at night, something skittered past my right eye. What was that? A while later, a second skittering. My curiosity was up. I turned toward the movement, lifted a chuck of old cholla, and there it was: a scorpion with its tail raised, daring me to come closer. Later I learned that "scorpion-ologists" also use the fluorescent trick to determine how many scorpions are in an area.

Joy, Terror, and Wonder in the Night

As an itinerant naturalist wandering across the continent, I have always been fascinated by the night. The time from "can't see to can see" holds a special place in my heart and also a special place on the corner of my bookshelf. When the sun sets, most naturalists retire to write, but a few—Vinson Brown, Lang Elliot, Diana Kappel-Smith, and Lorus and Margery Milne—explored the darkness and later chronicled their experiences in books like *Knowing the Outdoors in the Dark, A Guide to Night Sounds, Nightlife: Nature from Dusk to Dawn,* and *The World of Night.*

In the West, nights bring campfires, and campfires bring stories, of which I have a few to share. In stories, though, you don't tell about the

7.2. Bobcat

miles of stumbling along the trail, the hours of staring into the dark hoping to see something. Tales are made of brief glimpses into shadowed happenings. Most stories come from unusual occurrences and things that go bump in the night.

When I was fifteen, we had a predator call. When one blows the call it sounds like a wounded rabbit. That night my friend and I sat on a rock on the side of Laramie Peak in Wyoming. Mournful shrieks as we blew the call brought a pair of round yellow-white eyes advancing at about two feet above the ground. The mammal approached without a sound, and our hearts leapt. Finally, our flashlights revealed a large, healthy bobcat— its size said male—coming straight for us. Fear embraced me. We froze. What to do? We had no weapons. Suddenly the eyeshine was gone. Where was the bobcat? We sat for the next hour not breathing, not daring to. Would it attack? It didn't.

7.3. Yellowstone grizzly bear—not to be confused with camping gear

Then there was the night of my nose-to-nose run-in with a grizzly bear off the southeast corner of Yellowstone National Park. We had returned to camp that afternoon to discover our food had been ravaged by a grizzly. Claw marks were evident on torn-open stuff sacks, and our few cans of food were punctured by teeth marks. We cleaned up, talked late, and finally settled into a fitful sleep.

Sometime in the night, when—according to my storytelling partner, Jim Garry—"even the watchdogs sleep," I heard the first noise. It was near the head of my sleeping bag. I listened, trying to pretend I hadn't really heard anything, but it didn't stop. No denying it! I rolled to my stomach and stared straight ahead. My dark-adapted eyes made out a dim outline of a muzzle with light hair along the jaw and a massive body. Maybe ten feet separates us.

I froze. The dark shape froze. Neither of us moved. I don't know how long I wrestled with fear, but it was a stare-down with high stakes! When the bear didn't move, I realized I would have to break the stalemate. I could see my headlamp two feet off to the side. Slowly I inched my hand toward it, and finally my fingers clutched it. I decided to shine the light directly in the bear's eyes, roll right, and exit my sleeping bag to a position behind a tree. Heart beating, I flicked the light switch on, rolled, spun, and hid behind the tree.

The bear didn't move. I shone the light in its face, and then I realized that what I had thought was a bear face was my sleeping bag stuff sack hanging over a branch, where I had left it. The light-colored muzzle was the white strap in the middle of the bag. Even dark-adapted eyes can lie sometimes.

Living in the outdoors, being a biologist, and being a night scientist require skills not taught in college. In fact, these skills are rarely taught anywhere—skills such as not looking straight at faint objects so they will show up; letting your eyes get used to the dark; determining latitude, day of the year, time of the night, and passing of time; and spending an unexpected night out. At the National Outdoor Leadership School, I taught these skills to budding outdoor educators.

We were in the Sweetwater Desert, southeast of the Wind River Range in Wyoming. After three nights of camping in the river valley, just before "can't see," I would lead my students on the mile walk back to the cars to get something. Reaching the cars in waning light was no problem. But when the students turned to go back across the flat featureless desert, fear struck. They were lost. Worse yet, I would tell them they were on their own to get back to the tent camp that night.

After a period of self-questioning during which my students wondered why they didn't have warm clothing or matches with them, I'd ask them if they remembered my pointing out Orion the night before. When they faintly answered yes, I'd ask where the guide constellation was in reference to camp at this time of night. Eventually all would agree where Betelgeuse was, and we headed south back to camp, a bit wiser—and a bit more confident.

The human eye adapts well to the dark, far better than most expect. If you have a half hour for adaptation, a moonlit night becomes bright, and full moons cast visible shadows. The center of the human eye lacks light-sensing cells. If we look straight at the trail, we don't see it, but if we cast a sideways glance at it, the light hits the sensitive parts of the eye, revealing the trail.

Light shining on ski trails does strange things. It was 30°F below zero, and I had taken my biology class somewhere southeast of Heart Lake in Yellowstone National Park. This was a two-week biology ski course, crossing one hundred miles of Yellowstone in the winter wild. That night we were skiing for the fun of it and had ventured a couple of miles from camp. The time was right for a light lesson. I ordered lights out—no headlamps or flashlights. As we stood still, our eyes grew accustomed to the dark, and we were able to see more and more stars. Then, to my students' amazement, the ski trails stood out brighter than the snow around them. Light concentrates in a ski track, reflecting from side to side, thereby revealing the trail's presence to the night traveler. By focusing a little to the side of the trail, we could ski to a plethora of sensual experiences: "quiet you almost could hear," "stars so brilliant they hurt," "sound so crisp the swooshing of skis crackled," cold that "burned like a red hot spit," and "if our eyes were closed, then the lashes froze till sometimes we couldn't see." Robert Service and his ballads of the north hovered over us that night, instilling feelings we will never forget.

In the Unforgiving Land of the Noon Moon

Far to the north (for that matter, far to the south also) is a place known as the land of the noon moon. Daylight-centric people have long called this the land of the midnight sun. In polar lands, not only can the moon be seen at noon, it also circles completely through our visible sky every twenty-four hours. Wait six hours and the moon will be to your right. Six more hours and it will be behind you.

7.4. Near full moon

The land of the noon moon can be unforgiving of careless nocturnal biologists. My students and I were studying polar bears on the Arctic fringes of Hudson Bay. After supper, we decided to drive to town. As we rounded a tight corner on the dirt road, a flock of willow ptarmigans flew directly in front of our van—thud, thud, thud. I got out and found two unfortunate dead birds; a third bird had managed to avoid disaster. Figuring we'd eat them later, I tossed one in the van; the other was bloody, so I hung it on a willow bush, thinking we would retrieve it later when we returned from town and the blood was frozen.

Later that night, as the temperature dropped to 20° below zero, we returned along the road and I stopped at what I thought was our willow bush "refrigerator." Not spotting the bird in my headlights, I hopped out to check. The moon lit the night snow and cast shadows in the underbrush. The snow was over a foot deep, and I moved along the easiest route I could find among the willows, with a couple of

7.5. Moonrise over iceberg

eager students behind me. We spotted a drop of blood, and as I looked closely at the snow to gain perspective, it dawned on me, like the sun rising, that we had been walking pretty easily through the snowpack because we'd been walking in existing tracks—big tracks—fresh polar bear tracks! As adrenalin coursed through our veins, we quickly backed out, leaving our supper for our large white friend on the white snow under the moonlit night.

Later, my students and I lay pinned to the ground by the northern lights, awed by their incomparable beauty. The lights shone so brightly, we could read our field notes. Shadows became visible on the ground as green, yellow, and red curtains danced above. Curtains of light swept from the northern horizon to the southern in a matter of minutes. Overhead, the light rays converged on a single point, creating an auroral crown of the heavens.

Recalling our close call at the willow bush, I was determined not to make the same mistake. Beside me lay my twelve-gauge shotgun. While my students audibly "oohed" and "aahed" at the aurora, my ears

7.6. Polar bear on tundra

7.7. Aurora Borealis

were tuned to detect any movement in the shrubs or footfall on the ground. But no polar bears came our way that night—I like to think they too were watching the aurora.

Hippo Standoff in Africa

Some of my nighttime research projects took place in warmer climes and other distant continents. A tracking project took me and several students to Tanzania in 1976. Our plan was to track elephants, rhinos, and hippos as surrogates for understanding dinosaur trackways in North America and to interpret ecological relationships, such as predator-prey ratios between dinosaurian herbivores and their predators. Ultimately, we tracked everything we located—from dickey birds to lions to zebras and anything else that moved.

Our task was to show that we could look at tracks on our study plots in daylight and correctly discern the number of species and how many individuals had crossed the plot each night. Under the stars, well before daylight, I'd take my position on a small knoll and use binoculars to count animals and identify species roaming through the plots. It was exciting when the first lion crossed, the first giraffe, the first baboon. Most were simply passing through without stopping, presumably hunting or moving to feeding grounds or water.

One morning I spotted hippos. Although hippos were always along the river to the south and occasionally in the lake to the east, we at times found single hippo tracks across our plots. That night, however, it looked like two were coming from the lake. In the dark they looked huge (maybe males), but then so did my sleeping bag stuff sack in Yellowstone. I watched their odd movements: together, apart, together, apart, together, apart. Something was happening, but in the dark and at a distance I couldn't tell what.

Eagerly, I awaited dawn to see what had transpired between the two hippos. Finally, it was light, and animals scurried to their daytime haunts, mostly in the forest to the west. I ran to my plots, and there the tracks revealed a story. I saw two sets: one made by a large male and the second by a slightly smaller male. The tracks indicated that the animals had stopped and apparently glared at each other. Then the race was on. The two had loped forward (hippos can't hit a full gallop) and then began banging into each other's sides. The outside feet of each animal had pushed and slid out in the mud, first perpendicular to the track, then backward parallel to the track. The hippos had been side to side, each trying to push his rival away. Once, twice, they had smashed into each other. On the third collision of these colossal giants, the smaller one had slipped and gone down. Blood spatters showed that the tusks of one had made their mark. The two had then parted and faced each other again. The bigger one had turned and defecated toward the smaller one. The thick tail of the hippo swung through the falling scat, scattering it across the flats. It was as if the larger hippo had said, *Get the shit out of my territory!* The smaller hippo's tracks led into the forest, not a normal place for a hippo. The larger hippo had returned to the lake, triumphant.

Tale of an Arabian Night in Africa

Even nocturnal biologists need to relax some nights. My students and I had just spent two weeks crossing the veldt of southwestern Kenya on foot. Late that afternoon, we got to our vehicles and started for Nairobi. However, it was late, so we pulled off the road and set up camp before dark. We lit fires and began preparing supper.

Suddenly we heard shots, and looking into a dust cloud, we saw three approaching Land Rovers carrying men armed with guns. Knowing there had been poaching incidents and unexplained deaths in Kenya, I was gripped with fear. When the Rovers stopped, my fear abated somewhat. The men, who wore expensive white turbans, did not appear hostile. One leaned out of his vehicle and said hello, his accent British English. He offered our group fresh meat from animals the men had recently shot. My students, who had been on the trail for two weeks with no fresh rations and very little meat, looked at me hopefully, their eyes saying, *Yes, yes.* Our only meat had been a gazelle we had shot and smoked about three weeks earlier and that was long consumed.

Each student cook group was given either a guinea fowl (like a large turkey), a dik-dik (like a small pronghorn), or a spring hare (like a large jackrabbit). Skins were ripped off and flesh plopped in the cooking pots. The last bites of dinner were nearly down our throats when an old beat-up Chevy station wagon pulled into camp. Out stepped a Kikuyu native, who approached us and said in English that "His Highness" requested our presence at dinner. With a surprised laugh, I asked him who he thought he was kidding. The native tried to convince me that his invitation was indeed valid, clearly becoming insulted when I wouldn't go with him. I don't know if it was pity or curiosity that finally prompted me to say yes.

We jumped into our old Volkswagen buses and followed him into the desert twilight. After a mile across the sands, we peered over a rim into the Ewaso Ngiro River. Below us were many elaborate harem tents and rows of Land Rovers and Rolls Royces. Hanging from the trees were hosts of dead mammals and birds, the results of a serious hunting expedition. We drove to the largest tent—as long as a basketball court. Ushered inside, we could see tables laden with food at the back half of the tent. On one table was a large crystalline bowl full of peaches, pears, and mangos and wine and ice. Our eyes popped. *Ice*! We'd barely seen clean water in two weeks. Our drinking water in camp came from a hole dug in the mud. Water would seep in and we would scoop it up with a cup and pour it through our handkerchiefs into our water bottles. Next we added halozone—a decontaminant— and let the water sit for thirty minutes. Finally we squeezed lime juice into the water to help kill anything left alive—and the awful taste. So imagine how we reacted to the presence of ice! (Later, we learned that the ice had been trucked in that day from Nairobi.) Some of my students passed up the water and filled their glasses with wine. Those who did not want wine were presented with hard liquor, an expensive commodity in Africa.

About twenty minutes later, a group of men attired in fine white robes and white turbans entered the tent. Once again, we gaped. They were so clean! For two weeks, we hadn't had enough water to drink, let alone bathe or do laundry. We were flabbergasted when each of us was approached by one of the men offering to act as our personal host. My host was the president of the Kuwaiti branch of the World Bank. The entourage, we learned, was the royal court of the emir of Kuwait. Each year the group took a one-month vacation, and that year (1976) the holiday was an African safari—at a cost of $5,000 a day.

For the next twenty minutes, we all indulged in more drinking. Then in walked a stalwart man flanked by two very large gorilla-sized men we knew we didn't want to cross. I recognized the one in the middle. He was the man in the Land Rover who had said hello and offered us the game animals. He was, in fact, the emir of Kuwait. He took a seat across the tent from us and spoke in his native language. An interpreter informed us that it was the custom for guests to start the entertainment and that the emir had asked us to sing a song. We hemmed, we hawed. Then we did it! We sang the only song we could all think of that we all knew—"Home on the Range." We were a hit. The Kuwaitis loved it; they were on their feet clapping. They reciprocated by singing in their native language. As we all drank more, we were asked to sing again, but my memory fails me and I cannot now recall which songs we performed—the Kuwaitis were very generous with the drink.

After two hours of drinking and singing, one of our hosts announced that supper was ready. Remember, we had already pigged out at our camp. We joined the line forming at the table. My host pushed me to the front, and before me lay a smorgasbord of fresh meats, fine pastas, and breads. The first item was a whole guinea fowl. Not knowing what to do, I grabbed a leg and cranked it off. From the corner of my eye, I could see my host frowning. He plopped the whole guinea fowl on my plate. It hung over the edges. The next item was rice pilaf in a bowl—no spoon. I stalled. My host dumped the whole bowlful on my plate and the bowl was refilled. The next item was an Indian chapati ten inches in diameter. I looked at my host and knew to take the whole thing. It was replaced.

We sat. We ate. We drank. As soon as our plates were partly empty, they were filled with more food. As soon as our glasses were half empty, they were refilled. For two hours, we talked, we indulged, and—did I mention this?—we drank.

About midnight, we moved out of the tent onto the sandy river-bank, where a fire had been built. We drank some more. Our hosts, men who grew up in harem tents where there were no TVs and no movies, began a series of what you might call party games—games of dexterity which, after four hours of drinking, my students and I found quite challenging. Our hosts were amused by our attempts. I kept up our end with my repertoire of rope games and parlor tricks.

After an hour by the river (five hours of drinking now, if you're counting), the emir stuck a broomstick in the sand, and about thirty feet away he placed three bottles of hard liquor. A host got up, approached the broomstick and, bent at the waist, put his forehead on the pole. Then, as fast as he could, he ran around the pole ten times, all the while keeping his head on the broomstick. As he ran, the Kuwaitis counted his rotations out loud. (Consider what spinning around a broomstick ten times after five hours of drinking does to one's equilibrium.) When the host then attempted to run for the three bottles of liquor, he fell over, plowing into the ground with his nose. Another host, another nose plow, and so on.

Then the emir got up, pointed to me, and said "You"—his only English since "Hello." As leader of our group, keeper of our honor, I stood proudly and walked with dignity to the pole. I ran around it nine times, locking my eyes on those three bottles of liquor as I was finishing the tenth rotation. I kept my eyes locked on them as I ran toward them. And I didn't fall down, really I didn't. What happened was the earth turned ninety degrees and hit me in the side of the face. I swear to this day I did not fall down. The earth turned!

We had reached the sixth hour of alcohol consumption when the natives served us tea from a three-foot-tall silver teapot they had placed in the fire. Afterward, sometime between 3:00 and 4:00 in the morning, we drove off into the sunrise back to camp, never to see the emir of Kuwait or his court again. The next day I told my students, most of whom had severe headaches, that we always ended field trips with final events like the previous evening. I doubt they believed me.

A Bear of a Night

Great as Africa is, it lacks bears, a major disappointment for an urso-phile (one who loves bears) like me. Night and bears go together. Our ursine friends love to roam at night, especially during the fall when they undergo a chemical change known as hyperphagia, during which they eat anything and everything to put on fat for winter. On one of those fall days in Terrace, British Columbia, we were searching for white-phase black bears, known as Spirit bears by the natives. Sadly, we did not find Spirit bears that night, but later found more than we bargained for.

As evening came on, we hiked into a garbage dump, closed for the day, which local gossip indicated was often visited by Spirit bears. The dump was located on an embankment high above the river. There, among the piles of garbage, were lots of bear tracks, most made by black bears and a grizzly or two. Maybe one set was made by a white-phase black bear.

We took up positions on an excavated mound about ten feet high. Two or three black bears came in, but their appearance and activities were not remarkable in any way. We decided we would wait until dusk and leave if nothing more happened.

Time passed and daylight faded. We decided to leave while we could still just about see where to place our feet and had begun to pack up, when suddenly a "biiiiiggg" black bear came up the road. A very big bear. We decided to wait longer. The bear headed toward the garbage, and we started to get up to leave. Then stopped. Another "biiiiiggg" black bear was coming up the road. But bears look bigger in the dark, we reasoned, and we decided to let it pass our exit route on its way to the garbage.

Again we got to our feet. And again—you guessed it—another bear appeared. Now it was pitch dark, but my failing flashlight revealed another black bear. Wait. Then another bear. More waiting. Then more bears. More waiting. It was now approaching the bewitching hour. We decided to make a run for it.

Grabbing our packs, we slid down from the mound and glanced back. Bears were behind but not in front of us. One student couldn't walk quickly, but we hurried toward the gate as fast as we could. Rounding the first corner, I looked down at the ground and saw alternating right and left wet spots. Getting to my knees, I checked. They were bear tracks but much larger than we expected. We double-checked, looking more closely at them: the tracks showed long claws and webbed toes. Black bear claws are short, and the webbing does not extend far forward between the toes. Suddenly, we understood: these tracks were made by a grizzly, not a black bear. The tracks were still dripping wet from the river. But where was the bear? As if on cue from a conductor, we all broke out in unison, whistling, talking, singing. We walked out of the dump, positioned back-to-back in pairs, watching for bears in front of us and bears behind us until we reached our car and made our getaway.

Bat Nets and Owls

The consummate creature of the night is the bat. Like shrews on wings, bats dominate the night skies, preying on any insect that ventures forth. To capture and study them, mammalogists use mist nets, fine black nets that are hardly visible when stretched between two poles and sometimes not detectable by bat sonar until it is too late.

When I was a graduate student, my advisor sent me as part of a team to Canyonlands National Park to conduct mammal surveys. Our task was to catch as many bat species as possible by stretching nets over the water holes. However, we hadn't had much luck. With the field trip nearly over, I joined the park staff for a twilight game of volleyball. The superintendent stepped up to serve. There was a sharp *thwack* as his hand hit the ball, then a muffled *thump* as something fluttered to the ground. We descended on a brown mass smashed flat by its contact with the volleyball. Good scientist that I am, I collected the bat to stuff for a museum specimen, although as flat as it was, it could have passed for a herbarium speciman.

The next day, feeling that my bat-snagging luck might have turned, I set out again with my crew. Toward evening, we dropped down into a water hole surrounded on three sides by cliffs. There we strung nets across the water and waited for dark. Soon we could sense bats dive-bombing around our heads, and occasionally we saw them against the sky, but none were hitting our nets.

Frustrated, we waited, vowing to bring the volleyball the next night. Then one of the nets "jumped." Something had flown into it. The net bulged far to the side, farther than any bat would have caused. Suddenly, the net was jumping faster than drops of water on a hot stove. Whatever we had captured was in the center of the net, but the center was over the water hole, which was too deep for us to wade in. We tried to untie both ends of the net without falling in, our efforts hampered because each of us was blinded by the others' headlamps shining across the water hole. One of us fell into the water; then another. Finally, we had the net between us, but it started to spin out of control.

We worked our way hand over hand from each end toward the center, but as we got closer we could hear a loud distinctive clacking sound. It was not the sound of a bat, but we knew the sound meant business and sensed that our fingers should not end up at the spot where the sound was emanating from. Cautiously, we secured the net and finally got a flashlight on our captive. It was an owl, probably not more than twelve inches tall.

I scanned the owl for clues I could use to identify it. (I am not a birder.) Its dark eyes stared back at me—a clue. White streaks ran down the feathers of its chest—another clue. We later identified the bird as a barred owl. Barred owls are rare in the desert, making this an interesting capture. But despite our efforts, we captured no bats that night.

Tracking Wolves, Mice, and a Packrat in a Laundry

In Vietnam, I used night vision scopes to track enemy forces and often thought about the possibility of using the same technology to

7.8. Yellowstone wolf

track animals at night. By the 1990s, cheap first-generation versions of night-vision scopes were available in the United States, but at distances the results were not too pleasing. In 1977, when the technology had reached the third generation, a friend of mine, Michael Sanders,

7.9. Moonrise, Yellowstone

happened by chance to make contact with an employee from ITT. As a result of their conversations, ITT loaned us fifteen $4,000 units to observe wolves in Yellowstone National Park. Each unit had a light amplifier and a four-power scope.

As darkness descended on the first evening of our field trip in the park, I aimed my spotting scope at a sleeping wolf. Then I put the amplifier to my eye and looked in the scope. Wow! It was like daylight, but with a green tint. The results were far better than any we had in 'Nam. The others in the group trained their units on the wolves and, like magic, a host of behaviors we had never dreamed of was revealed. Eventually, the wolves loped off into the woods on a hunt, ending our observations for the night.

With the departure of the wolves, we turned our amplifiers on the stars. A clear night in the light-pollution-free wilds of Wyoming is mind boggling, with the Milky Way painting a continuous band of light across the sky and stars too numerous to count. Words cannot describe what becomes available to view through a night scope. Constellations stand out prominently from their competing solar neighbors. Stars and nebulae litter the sky and jump out at us. That night, distracted biologists became astronomers, failing to notice if the wolves returned.

Our success with night vision scopes whetted my desire to try to observe the nighttime movements of small mammals like mice, so I invented a "light collar," which consisted of a small chemical light tube called a light stick. By carefully cutting off the top of the translucent rubber tube and pouring the chemical into a small jar, I could extract the small glass vial from inside the tube, open it using a file, and pour the contents into a second jar. The collar to go around the animal's neck consisted of a neck band made from a nylon tie and a capsule made of clear rubber surgical tubing, which I sealed. After the mice and voles were trapped, we would inject the two saved liquids into the hollow tubing using syringes. Presto: a light-producing collar. We placed a collar around each animal's neck and snugged it up so it wouldn't fall off. The mice were then turned loose, and we followed them, making notes.

In the 1980s, we trapped mice at the University of Colorado Mountain Research Station's main laboratory, located above 8,000 feet. Each mouse was ear-tagged, and then we'd move them farther and farther from the building and release them. The homing skills of mice are excellent, and soon they'd be back. In fact, every one of them made it back. So we kept moving them farther away.

Finally, we took the mice across a fast-flowing creek, decidedly too wide for a mouse to jump and a half mile from the main laboratory. One mouse almost beat us back to his peanut butter reward, but in the live trap. Surely, this mouse had never been to where we had dropped it off, and how, we wondered, did it cross the creek?

Out came the light collar. On the next night, we collared our mouse and returned it to the drop point. It was pitch dark. By the light of our flashlights, I shook the trap upside down and out popped a glowing ball of light. We backed away. The light ran in circles for maybe a minute and then turned directly toward the laboratory. We all followed at a distance, mesmerized.

When the mouse reached the stream, we held our breath. Would it swim? Our mouse paused . . . well, we saw the light pause; we couldn't see the mouse. (We were careful not to direct our light on the mouse so as not to influence its travels.)

Then, to our surprise, the light rose vertically in the air. Even scientists give in to curiosity: I momentarily turned on my flashlight for a quick glance at the mouse. My beam revealed the mouse climbing up the trunk of a pine tree. We watched. Ten feet up, fifteen feet up, then the mouse moved horizontally in the direction of the lab. Suddenly, the light seemed to fly, dropping from about fifteen feet to ten, but now it was on the *other* side of the stream. The light descended the tree while we waded across. We followed the light and the mouse—right back to the peanut butter in the lab.

With the coming of daylight, we were out examining the crossing point. The tips of the pines extended most of the way across the stream, but there was a gap. In the dark of the night, that mouse jumped the gap. Was this the route he had used to return earlier? How did the mouse find trees bridging the stream at night? How did the mouse see when our eyes didn't? Herein lies a master's thesis.

Although bats are the consummate mammals of the night and mice seem to be able to jump from tree to tree in the dark, packrats are equally efficient denizens of dark, as this next adventure demonstrates.

I arrived at Pine Butte Swamp Preserve in northwestern Montana to teach grizzly-bear tracking, as I often did. The preserve, an old dude ranch, is owned by the Nature Conservancy, which accepts guests to augment its conservation budget. Ranch manager Lee Baraugh was at his wits' end. Packrats were everywhere, and he couldn't get rid of them. Lee solicited my help.

When I went to my cabin, I knew immediately that Lee had problems. Packrats have a distinctive smell that I can pick up a mile away— well, maybe not that far, but I can tell when they are in a cabin, and

7.10. Cabin at night

they were indeed in this one. Packrats also have a habit of drumming a front foot on surfaces when they are nervous. Walking around the cabin, I could hear the telltale drumming. Without doubt, rats were there, waiting for me to go to sleep to venture out of hiding. I could have set my traps and killed one or more, but . . . well . . . they're cute. So instead I drifted off to sleep and dreamed a plan for Lee. Next morning, I told him yes, we could kill a few packrats, but more would simply move in. I convinced Lee that we had to determine their travel routes so we could deter newcomers. (I really wanted to study how the packrats were living in the complex, but I didn't tell Lee that.)

Out came the fluorescent powder. Powder, packrats, and Murphy go together. (You know Murphy's Law.) I set traps that night, and in the morning a fine specimen of a female was in the trap. I shook her inside a plastic bag containing the powder and let her go, to the delight of my students.

We allowed her to run around undisturbed for a night, then on the second night, about midnight, I brought out my ultraviolet light. We were off like a pack of hounds on the trail of a fox. The packrat exited the cabin, crossed the grassy yard, and entered another log cabin. As we reached for the door, a glowing rat emerged from under the eave. On the up side of a log, the rat ran around the building, followed by an eager entourage of student scientists. Down and across the lawn it went, entering the utility cabin by squeezing under the eave. Then it disappeared.

Holding our breath, we approached the door. Silently, we breached a gap and saw tracks going down the hall to the laundry room. On tiptoes, we crept in and opened the laundry room door. It glowed! There was powder on the floor. There was powder on the washing machines. There was powder on the walls. There was powder on the ceiling. There was powder on the towels, the washcloths, the sheets, the guests' laundry. There was powder on the soap boxes. We're not talking about a speck or two of powder here and there but rather copious glowing amounts sticking to nearly every square inch of the room.

The tracks led to a laundry basket and there, nestled in the cook's clean underwear, lay the mother packrat and her nursing babies, staring up at me. Mom froze, not twitching a whisker—tough task for a packrat. My agreement with Lee was to kill all packrats. My students were at my back. I could feel twenty eyes on me. This mob was prepared to attack if I made the wrong move.

Time seemed to stand still as I weighed the conflicting ethical demands I faced: I had given my word . . . but those eyes, those big innocent eyes (have you ever seen a packrat's eyes?). "Shhhhhhh," I said quietly, and I backed out of the room, past the students.

In broad daylight, specks of fluorescent powder will show on white laundry even without a black light. However, we weren't dealing with specks—we had gobs. I knew the proverbial odoriferous material would hit the air-circulating mechanism come morning. The maids

7.11. Devils Tower, Wyoming

and the cook would go ballistic. They would never have known that Mama had walked all over the laundry if it weren't for the copiously distributed powder. Worse yet, in camp there is one person you never want to upset—the cook. And I'd be standing there without a skin to show for my trapping efforts.

I dismissed my students with orders not to disturb our nursing mother. About the time when even the watchdogs are sleeping (heard that one before, but it was a safe time of night), I crept into the laundry room with a live trap. Two hours later a metallic twang told me the trap had shut. Inside I found Mom, savoring peanut butter. Quickly, I grabbed her nest and the pups in it. At the far end of the pasture, past the horses, was an outhouse with its door nailed shut. I knew squirrels had a large soft midden inside. I opened the door with a hammer and deposited the nest and then Mom into a hollow I burrowed in the midden. I nailed the door shut and was off at dawn of day to meet Lee.

Now Lee was a cowboy ranch manager dead set on eradicating rats, but likely he never visited that old outhouse again. Mom and young were safe there.

I told Lee I'd killed that damn packrat, but there had been a bit of an accident with the powder. Apologizing, I explained that everything in the laundry would have to be washed because of the spill—failing to mention anything about the packrat having been there. Now, do you know how hard it is to get hydrophobic powder out of laundry, underwear, and washing machines? Well, that's a story in itself.

A Devil of an Experience

I didn't mean for this piece to be a treatise on packrats, but they do seem to pop up frequently in my night adventures. In this exploit, a packrat literally popped up and provided valuable advice at a critical moment after fellow graduate students signed on to one of my schemes. This story began when Dr. Ruth Bernstein had excited us all about island biogeography. The first challenge to an exploration of the subject was our location—in the center of the continental United States, where oceans are hard to come by. I reasoned that the top of 1,000-foot-tall Devils Tower National Monument in Wyoming could serve as an island, with its steep sides separating the flora and fauna of the top from the bottom. It was easier to convince Ruth of the value of this plan than the superintendent of the monument, but eventually permission arrived.

Ruth was to lead a group of graduate students in collecting every living thing that was collectable around the base of the Tower. Two of my fellow graduate students, Joe Beckman and Charlie Fuenzalida, and I would climb to the top and collect there. To do a good job of collecting, we would need four days on the top of the Tower, so we'd have to bivouac for three nights. When we set off at dawn on a May morning, we were carrying three one-hundred-pound duffle bags full of water, food, sleeping bags, traps, and other collecting paraphernalia.

We were the first climbers on the Tower that year. Winter ice was still deep in the cracks. Even in good weather, climbing the Tower is far from easy. The climb was a "beat-out" from 6:00 a.m. to midnight before, exhausted, we reached the Meadows two-thirds of the way up. The Meadows tilt slightly away from vertical, and there is a pocket where you can squeeze in with a sleeping bag. We collapsed there for our first night on the Tower.

In the morning, we made the easy climb of the final third and we were on top. We set out traps for small mammals, collected ants for Ruth, and gathered plants for others. As light faded, we could see dark clouds to the west. After fixing ropes so we could easily come and go from the top, we retired to the Meadows for our second night's bivouac.

Night on the side of a 1,000-foot rock face is boring unless you're a biologist, a night biologist. In fact, our location was a night biologist's behavioral dream—at least for those interested in deer mice. When the sun set, the mice came out. Amazingly, the mice on the Tower have no fear of people, perhaps because of their long isolation. They scrambled up and down the vertical rock walls of the Tower. They came into our camp, searching for a scrap of food, perhaps curious about the first humans they had seen on the Tower at night. They hopped up on our sleeping bags and looked directly in our eyes. Apparent territorial battles ensued with chases and counter chases. The mice became our entertainment for the evening—better than television. Even when we drifted off to sleep, they scurried across our faces, waking us. (Occasionally packrats would come to the edge of our lean-to cave and peer in. Their smell had earlier revealed their presence but they lacked the boldness of the mice and did not come close.) By 8:00 that evening, it was raining. By 9:00, it was slushing. By 10:00, it was a full-scale blizzard. On the Meadows, a slab of rock slopes away from the vertical wall, under which there is room for two sleeping bags, but we had three. Water started to come in from both sides. Wind blew, snow flew; by morning, all was white, wet, and icy. That night, as the blizzard raged, mice took advantage of our nook. Once curious about their human guests, they now seemed to search for food oblivious to our presence as it snowed and dripped.

By morning I could see that sleeping bags were getting wet. Charlie had a feather bag even though I had told him to bring a synthetic one. The feathers were wet and starting to clump. There were still traps on top of the Tower with animals in them, so we fought our way back to the top and collected mice and packrats. It was treacherous going, to say the least. We did what work we could and dropped back to our hanging camp. The blizzard was in full force, and the day continued blustery, miserable, wet, and worse, but we hung in. Simply put: there was no way off the 700-foot face below us.

Night 3 was serious. Mice no longer entertained our thoughts. I could feel the others trembling much of the night. They had not brought all the wool and synthetic materials they were instructed to. How long would it be before hypothermia set in? There would be no rescue. The adventure seemed to be turning into an epic—possibly a tragedy.

I woke in the dark to mice scurrying over me. The night was a haze in my mind: had I been dreaming of a blizzardy world on the side of a mountain? I slept again, then woke to the strong smell of a pack-rat. Rolling over, I heard the telltale tapping of a front foot. Curious, I turned on my flashlight and sure enough, sitting there alongside my bag was a packrat. It looked me in the eye and said: "Mom told me to say thank you for Pine Butte and to get your butts off this mountain before it gets worse." I blinked. I was awake. Was I awake? To this day I don't know!

By daylight, the situation was serious. We had to get off the mountain. The falling snow had dwindled, but the wind was blowing hard. I considered leaving behind the traps we had on top, but we needed the ropes we had up there to get down. Steeling ourselves and bundled as best as possible, we climbed to the top. Wind shoved us around, keeping us from venturing near the edge. The wind had scoured the high spots, depositing snow in the hollows where our traps were. We found some of the traps, but the high winds must have blown others off the top.

I belayed Charlie and Joe down to the Meadows from the top. Then I had to lead down without a top belay for safety. When we were finally at the Meadows, it was getting late. I wondered if we could make it down, but another night might have a devastating and perhaps lethal effect. Would this turn into a life or death struggle?

To the west an opening appeared in the clouds and the setting sun shone through. It created a rainbow. From our advantage point 700 feet in the air, the rainbow looked like it was on its side, forming a letter "C"! I still have a photo of it. The rainbow was our cue: go now during the lull.

We tied our 200-foot ropes together and dropped the gear to a ledge below, and then we rappelled in stages until we hit the bottom. No one was there to greet us. In fact, no one was even worried; their comprehension of how our life and death balanced on the Tower's face was nil. Only the packrat really knew.

EIGHT

Volcanoes and Fruit Bats
Fear and Loafing on Montserrat

SCOTT C. PEDERSEN

The volcano had been grumbling for several hours, rolling great glowing boulders down the flanks of its steep slopes in my general direction. It was July 1997 and I was nearing the end of a very long night after a very long day. I tried not to take any of this too personally as the volcano was nearly two miles away. Still I was mesmerized. Standing there in the dark, engulfed by a near-deafening chorus of tree frogs, watching gigantic embers crashing down the mountainside, showers of sparks and debris marking each collision—I had a front-row seat at a private and very surreal fireworks display. My abject fascination with these sights and sounds was quite rudely interrupted by a great searing pain radiating up my arm from my hand. I had been careless, distracted from the task at hand: removing a muscular pig-nosed bat from one of my mist nets. The young male bat had opted to impress this fact upon me by latching its teeth into the flesh of my thumb.

Before I go on, let me provide some background for the events that led up to this rather painful vignette. The old tattered field notes

DOI: 10.5876/9781607322702:c08

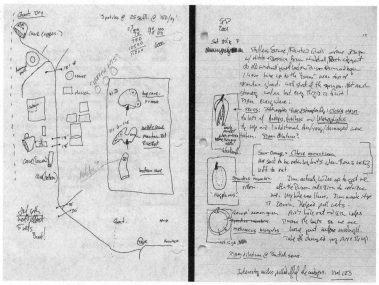

8.1. Author's field notes

covering my research on the bats of Montserrat, British West Indies, are a good place to start. The first couple of pages present a remarkable multilayered tapestry, replete with coffee-cup rings; subsequent pages are partially laminated together by what appear to be sweat rings left behind by beer bottles (undoubtedly Carib lager—the Beer of the Caribbean). Within the first few pages, I find a half dozen mosquito carcasses with blood and guts splayed around their mummified remains in bleak testament to their last moments and their last meal . . . me, if memory serves. Several business cards, tax stamps, expired driver's licenses, and peeled-off beer labels are stapled haphazardly along the page margins—the staples exhibiting a crusty patina of rust here and there. Today, my students tease me that my field notes resemble papier-mâché sculptures decorated with my unintelligible ink-blotchy Sanskrit. But my field notes are historical artifacts, testimony to my experiences on Montserrat.

A Night on a Bench

I made the first entry in my Montserrat field notes on August 22, 1993, the echoes of my dissertation defense still throbbing in my head like a mid-range hangover. Here I was, stuck between delayed flights on the island of Antigua, in the airport bar, scribbling away at my notes on three-hole 8.5 × 6.5-inch 100 percent cotton-rag paper with my trusty refillable Koh-I-Nor Rapidograph ink pen, trying very hard to make some sense of the previous couple of days. I was heading to the neighboring island of Montserrat to answer a cryptic two-line, sixteen-word job posting I had seen in Science—something about needing a gross anatomist in Montserrat, British West Indies. A fax number was included. I remember pieces of the subsequent phone interview with the academic dean of the medical school: "Do you have a pulse and a current U.S. passport? You're hired. Be here in three weeks." *Click.* I wasn't overly impressed, but hey, I was soon to be solvent! And, for the first time, independent. I was no longer a student under the direction of a professor: the decisions—and the mistakes—would all be mine.

Montserrat? Hell—I needed a powerful magnifying glass to find this flyspeck in my atlas of the world. Chris Columbus never even bothered to set foot on this rugged volcanic island in 1493, although he dubbed it Montserrat after mountains of the same name near a monastery back in Spain. The small (one hundred square kilometers) and extraordinarily beautiful island is located in the northern Lesser Antilles about 450 kilometers southeast of Puerto Rico. The British eventually colonized Montserrat in 1632 and had some luck growing sugarcane, cotton, and limes there. These were the limes that kept the British Navy from getting scurvy during long voyages, in turn giving these sailors their nickname—Limeys. Montserrat remains one of the very last British Crown colonies still in existence.

But here I was at the airport bar a few weeks after the fateful phone interview, with ink blotches all over my notes and fingers because my temperamental Rapidograph had not appreciated the decompression

8.2. Bottomless Ghaut, Montserrat

during our descent into Antigua and was now throwing a tantrum. Fortunately, a second and then a third beer proved to be quite neighborly, each in turn agreeing to accompany me out onto the veranda. I enjoyed watching two species of free-tailed bats (*molossids*) performing their nightly aerial ballet while filling their faces with seemingly invisible insect prey above parked aircraft on the hardstand, still sparkling from a rain squall that had passed through an hour before. Velvety Mastiff bats (*Molossus molossus*) and Brazilian free-tailed bats (*Tadarida brasiliensis*) are very common throughout the region, but they were new to a Nebraska boy. I was taken aback by the lack of more familiar critters—neither swifts nor swallows, just these free-tailed bats. I found myself musing that the singular ease of their movements—their speed and agility—made birds look terribly clumsy in comparison. And I think they were flying for the sheer joy of it. The acrobatic maneuvers clearly outnumbered insect interceptions—at least by my count. Perhaps they played at scaring the hell out of each other during their midair game of "chicken"—some sort of teenage chiropteran rite of passage. (In retrospect, perhaps this rather Hunter Thompsonesque moment was due in no small part to the fact that my brain was now happily floating and bubbling along in a fourth, maybe a fifth fizzy yellow beer—everything in my world appeared copacetic.) But, as happens all too often, reality reared its ugly head, snapping me out of my bats-and-beer reverie. The air-conditioned bar/restaurant locked its doors for the evening, and I didn't have accommodation lined up. That night I slept on a rickety open-air airport bench under the cartwheeling bats and the brightest stars that I had ever seen—painted in such thick brushstrokes that I suddenly felt embarrassed to think that I believed I'd seen the Milky Way before.

Mountain Chicken and Goat Water

When I finally made it to Montserrat, I set about moving into my new office. I soon learned that all of the electrical outlets readily accepted a standard 110V plug but, sadly, many outlets were actually wired for 220V. To this day I pity the poor radio–cassette player that met an untimely death because of my ignorance.

That first weekend, I set out to explore the nightlife in the woods adjacent to my home. Tree frogs, marine toads, night herons, large hand-sized moths, and small gray rats were common sights as I stumbled about in the bush that night. I heard a noise that I thought must be made by a frog, but I could never locate the critter, though its clucking-croaking sound seemed to come at me from all directions. Later that evening, I checked into a local watering hole and asked around about the elusive animal. The denizens of that well-worn rum shop informed me that this beast was in fact a "mountain chicken." A chicken? I was incredulous. I imitated the clucking noise to the best of my ability (much to the amusement of all), and everyone agreed that the sound came from an animal found well up into the mountains that did in fact taste like chicken—and what, I thought, doesn't? Turns out that the mountain chicken is actually a type of frog (*Leptodactyllus fallax*), and a really big one at that! Large females have a snout-vent (nose to butt) length of 210 millimeters (11 inches) with a stretched-out whole-body length of 450 to 500 millimeters (20 inches). Historically, the mountain chicken has been found on at least five major islands in the Lesser Antilles but is now restricted to the islands of Montserrat and Dominica. It is absent from the intervening French island of Guadeloupe, presumably because of the French penchant for eating the damnedest things.

During my third week on Montserrat, I met the clinical dean of the medical school, Nancy Heisel. My thirty-third birthday was approaching, so Nancy suggested we go out to one of the finer restaurants on the island to celebrate. It was a lovely little place, but much to my horror, the menu listed goat water and mountain chicken. I wasn't entirely sure that water associated with goats in any fashion was a particularly wise choice for a culinary offering, and I was having a very difficult time wrapping my brain around how exactly the chef was going to wrestle a gigantic frog that wouldn't stop clucking into his repertoire . . . or into his stew pot. Fortunately for me, Nancy (an infinitely competer practitioner of internal medicine) noted my dilemma and immediatel,

8.3. Mountain chicken stew

ordered up several (medicinal) beers for me. After the drinks arrived, Nancy engaged our waitress in a lively discussion about the menu: "Aren't mountain chickens protected? . . . I'm not sure we should be eating an endangered species. . . . Why are they on the menu?" Our well-coiffed young waitress politely interrupted Nancy to chirp, "Oh, that's okay, we only serve those that are already dead." Mind you, I was raised on Nebraskan syntax, but it was painfully unclear as to what we were about to be served—a road-kill frog, perhaps, or one that had died of natural causes a few days earlier? Or maybe there was some black-market amphibian abattoir just down the back alley that delivered fresh frog meat to this restaurant on the sly? We went with the lobster. After all, it was my birthday and my personal physician was buying. I was in paradise.

For inquisitive minds, goat water is a delicious savory stew. Family recipes are carefully guarded secrets on Montserrat, but I can reveal the basic ingredients.

2 1/4 lb (1 kilo) goat meat, cut into bite-sized pieces
2 large onions, sliced
2 tomatoes, sliced thickly
2 cloves garlic, chopped coarsely

3 whole cloves
2 tbsp. each butter and chili sauce
1 tbsp. flour
Salt, pepper, and Tabasco sauce to taste

Fig Tree from Hell

Because of my busy teaching schedule, it took me several months
to work my way over to the far side of the island and into the Paradise
Estate. This large abandoned estate was located on the windward side
of the island in a deep, lush valley laced with narrow roadworks laid
carefully across hand-hewn stone bridges that span trickling brooks,
bracketed by towering tree ferns, banana, heliconia, elephant ears,
hardwood trees, and the biggest damn fig tree I had ever seen. This
thing was huge! On February 26, 1994, I made the mistake of parking
my nice clean Honda beneath this evil arborescence. I had gotten a
late start that evening and only had time to set a pair of eighteen-foot
mist nets, one across the road forty feet beyond the car and the other
forty feet below the car. My Honda was my office; I rolled the windows
down, sat in the backseat, opened a beer, cracked my field notebook,
and started writing. I figured I'd catch a dozen bats over the next three
or four hours and call it a short night. I was dead wrong.

I was unaware there was a large rock overhang, just up the hill, out
of sight, that housed the island's gang of pig-nosed bats (*Brachyphylla
cavernarum*)—approximately 5,000 of them. This raucous species
alternated between this large cave-like structure and another very
large cave complex at the north end of the island, using each site alter-
nately as a regional bivouac as the colony tracked fruiting trees across
the island with the changing seasons. I was also unaware that the fig
tree from hell was at its peak fruit production, although a trained eye
would have noted the carpet of fallen figs on the ground *before* parking the car under this particular tree. Needless to say, the *Brachyphylla*
knew all about this tree and tackled it en masse as they exited their

roost. At first, the sound was deafening, sounding very much like a sudden tropical rainstorm. I quickly rolled up the windows so I wouldn't get soaked. Within minutes, I could no longer see out the car windows. But it wasn't rain—it was much, much thicker. I got out of the car only to realize, much to my horror, that both nets were on the ground, full of dozens of pissed-off, squawking, thrashing bats. The car (initially white) was now brown with bat feces, urine, fruit pulp, intact figs, half-eaten figs, one young bat, two tree frogs, leaves, small twigs, flowers, and God knows what else. As I slipped around in the muck in a vain effort to get the bats and my nets off the ground, I witnessed several midair collisions as these animals tried to get out of one another's way. Several startled bats even *thwapped* into me during the melee. It was spectacular!

I could do nothing for my poor nets—old Japanese things with cotton loops. They were done for. Given the squashed figs, twigs, leaves, snared bats, clouds of mosquitoes, gnats, and fruit flies, I had no choice but to shred the nets and toss most of the rather disgruntled bats back into the air. I did keep two dozen bats so that I could measure their body weight, forearm length, reproductive status, and so on, but released the rest as quickly as possible. Of the bats I kept, half were Jamaican fig-eating bats (*Artibeus jamaicensis*), very common throughout the Caribbean. Though similar in size to pig-nosed bats, these *Artibeus* were often chased away from feeding trees by the more aggressive *Brachyphylla*. Perhaps in this super-abundance of figs, the *Brachyphylla* didn't notice or care about these interlopers amid all the confusion of the evening. After an hour or so, the initial chaos had abated and I set up two more eighteen-foot nets, planning to monitor these two *very* carefully. With this level of bat activity, I didn't want to miss an opportunity to catch something really new, big, or interesting. Little did I know.

Given that one only "rents" beer, and there was a full moon, I strolled some distance away from my nets to appreciate both aspects of nature. As I came back up the road I heard a gentle human voice praying in

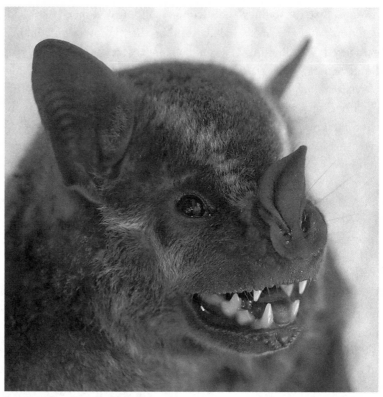

8.4. Jamaican fruit bat, Montserrat

the dark . . . very softly, as though he *really* meant it. And then I saw him: not over five feet tall or more than one hundred pounds, there he stood with a huge bundle of wattle strapped to his back with twine and vines—a makeshift bundle easily twice his size. This wizened old gnome and his great load were well tangled in my lower net. I killed my headlamp and tried my best to calm him down. I extracted him and his load from my net and sat him down on the road bank behind my car. I tried to explain what I was doing out here in the bush and had the presence of mind to offer him one of the beers I had on ice in the trunk of my car. He readily accepted the sweating bottle with thanks, held it in his knotted hand and rolled it slowly and purposefully across his forehead. Then he sniffed it appreciatively and downed it in a single draft. After taking a long deep breath, he explained in a most beautiful Montserratian patois that he had truly believed he had been snared in some gigantic spider's web and my bobbing headlamp coming back up the road signaled his imminent grisly demise. I had to turn my head away so that he could not see my smile.

We talked for an hour or so. He entertained me with wonderful stories, telling me the bush was full of many spirits at night, not the least of which was the rat-bat—a term I had heard frequently across Montserrat. After all, one never really sees rats and bats at the same time, so it is logical that rats might very well turn into bats at night, and then turn back into rats at daybreak. Although he didn't claim to have seen the transmogrification himself, he was nevertheless quite convinced that it happens when no one is looking.

After killing two more of my beers, he hit me up for some money to "give to the Lord" the next morning during church service. All I had was a U.S. $20 bill that I kept stashed in my field bag for emergencies—it had been through the washer and dryer so many times it looked like tattered facial tissue. For him, it was a minor fortune. Several weeks later, I learned through the grapevine of cabdrivers* that my ratty $20 bill had made quite a stir when it appeared in the offering plate . . . as did a certain woodcutter's story of the giant spider that wasn't.

---

* During my time on Montserrat, it became quite clear to me that if I needed to find anyone, needed to purchase a particular item (anything at all), or needed something done (correctly), all I had to do was go down to the taxi stand in Plymouth and talk to the cabdrivers. If all the super-computers in all the countries of the world were somehow networked together, they still could not compete with the blinding speed and accuracy of the cabdriver neural network on Montserrat, through which answers, products, and skilled craftsmen appear spontaneously as if conjured from the celestial ether. To this day, these drivers remain a vibrant and integral component of the fabric of this small island.

Early Rumblings

During one of my first lectures in gross anatomy, I noted that several students in the back of the room were looking quite agitated. Soon all 150 students were exchanging glances in astonishment. Then I too became painfully aware of an odor not unlike some great gastrointestinal discharge. This gaseous insult was followed by a small tremor—nothing too startling, mind you, but enough to get the fluorescent light fixtures in the lecture hall to chatter. This was the first of many sulfur-dioxide discharges and tremors I was to experience on Montserrat in 1993 and 1994.

When I moved back to the United States in 1994, I couldn't have known that the fate of this beautiful island had already been sealed. A year later, it came as a horrible shock to see Montserrat featured on CNN as the first thick clouds of gas and dust were emitted by the local Soufrière Hills volcano. At first, damage was limited to pollution of the streams by acid rain coming out of these deadly clouds. But the horror and heartbreak had just begun, and a great nightmare was just beginning.

What typically comes to mind when one hears about an erupting volcano are images of great glowing flowing ribbons of nasty red stuff, the sort of images one equates with pictures of the great volcanoes on Hawaii that grace the pages of *National Geographic*. But lava is kid's stuff. If you want real bone-chilling apocalyptic images, consider a pyroclastic flow—superheated (300°C) suffocating clouds of ash racing toward you at more than one hundred miles per hour, replete with heaving avalanches of huge glowing boulders. This was the geological horror that visited itself upon tiny Montserrat. Like some gigantic obscene amoeba, massive pyroclastic flows and mudslides (known as lahars) have slowly swallowed and entombed the island's trim capital of Plymouth.

I would never presume to try to paint an accurate picture of the human tragedy that beset this small island community. Official and unofficial reports of the number of people killed by volcanic activity

8.5. Soufrière Hills volcano, Montserrat

in 1997 differ, but the souls who died unofficially are dead nonetheless. Thousands emigrated to Antigua, Tortola, England, Canada, and the United States. Families were separated and much of the vibrant, unique culture that I briefly observed in 1993 and 1994 was spread upon the trade winds, an upheaval that has been termed the Montserratian diaspora. Those who were stuck on Montserrat early in the volcanic crisis were crammed like Spam in cans into churches, private residences, and emergency shelters for several years in the relative safety at the north end of the island in what can only be described as an abject social services disaster.

Belham River Valley Shake and Bake

I began this chapter by relating how the Soufrière Hills volcano had interrupted my work with an impertinent pig-nosed bat in the Belham

8.6. City Center, Plymouth, the capital city of Montserrat, after volcanic eruptions

River Valley in 1997. That particular evening, I had set my nets a bit farther up into the exclusion zone than I should have, but I had sampled the site before and I wanted replicate data. From my hillside perch, I

8.7. City Hall, Plymouth, buried under pyroclastic flows after volcanic
   eruptions

could easily see the abandoned hamlets of Weekes and Corkhill. I sat
there for hours, tending my nets, listening first to the tree frogs and
then to the barking of dogs. The barking was sporadic at first—dogs

working through their own lonely angst and quite probably abandoned to their own devices as these villages had been officially closed down by the government because of safety and liability issues. But the dogs remained. Alone. And then, it was as if the barking was orchestrated—I heard dogs well off in the distance at the upper end of the valley barking; then barking started up just across the deep ravine from me; then dogs farther down the valley joined in the chorus. That's when the earth shifted beneath me ever so very gently. Then everything went silent. Even the tree frogs shut up. Then another tremor chased the sound of barking dogs down the valley, and this sequence was repeated again and again.

I began to detect sounds far above me, so much like those made when a heavy truck rolls down a gravel driveway that I stood up to warn the oncoming driver of the nets I had stretched across this access road. I felt quite foolish when I realized there was no truck. My curiosity eventually got the better of me, and I hiked a kilometer farther up the valley, where the sounds were much clearer, more distinct, not unlike the *pawk-pawk* sounds bowling balls make when they clack together. Looking across the valley to the south, I could just barely make out against an inky black sky the ghostly presence of smoke, steam, and ash billowing up from the volcano's peak. Fiery red boulders rolled down the mountainside only to disappear in great showers of sparks and tumbling embers. (Time-lapse photographs would later depict similar events as red-orange-yellow streaks of material reminiscent of lava.) Needless to say, I beat feet back to my mist nets to extract a few bats—unremarkable save for the *Brachyphylla* that tried to eat my thumb. I pulled these nets in record time and began the long walk back to the valley floor, my car, and my last net.

That year, the entire Belham drainage was looking quite dry and threadbare. Pyroclastic flows had barbequed the houses and trees at the upper end of the valley in the hamlets of Molyneux and Dyers, and mudflows had begun to suffocate the river as it meandered across the valley floor. These events were clearly a setback for the Montserratian golfing community, as the delightfully quirky thirteen-hole golf course that had spread across the bottom of the valley was slowly being dissected and plastered over with thick layers of pumice, volcanic ash, and sand. This valley had also been one of my favorite netting sites back in 1994, as I had captured eight of Montserrat's ten species of bats there. On this particular night, I had placed a single mist net across the remains of the Belham River near my parked car on the off chance that I would catch something flying through in that damaged silt-choked habitat. Sure enough, hamstrung and hog-tied in that net was an oily, pungent, orange abomination—a fishing bat (*Noctilio leporinus*). Though capable of hawking large flying insects or snagging small fish from the ocean surf, fishing bats specialize in taking minnows, tadpoles, and insects from the surface of fresh-water streams and ponds like those that had been located along the floor of the Belham valley. But now this riparian habitat was effectively gone, and despite focused efforts to find it again, this capture in 1997 was to be my last sighting of the unique species until 2004. The seven-year hiatus left me wondering if this species had in fact been extirpated on Montserrat. How it survived is anyone's guess, but I surmise that the volcanic events of 1997 and 1998 were perhaps a very close call for this unique species.

Mudbugs and Mountain Chicken Redux

Sitting in the dark for hours, waiting for bats to hit your net, the mind wanders. However, some of my colleagues have spent this time more productively—hunting their dinner. James "Scriber" Daly and Phillemon "Pie" Murrain agreed to accompany me into the Farm River drainage to do some netting above the half-buried scorched remains of Blackbourne airport. Pie's father had been killed by pyroclastic flows in a village near this very site a year earlier. Later that night we were to toast, enthusiastically and repeatedly, the dearly departed. But for the time being, while I sat and tended my mist nets, Pie and Scriber were leap-frogging across mossy, slippery rocks up and down the Farm

River like two little kids, hollering and splashing, plunging their arms blindly under rock ledges. They returned to our camp with two dozen of the largest crayfish I have ever seen, skewered unceremoniously on a long branch they had used as a gig. These critters (*Macrobranchium carcinus*), quite different looking from the mudbugs I had grown up with, had chopstick-like pincers and an elongate spindly body. The largest specimens were at least two feet long. That night's collection was ultimately destined for a boiling pot in which the unhappy arthropods would involuntarily accompany some stewed onion, garlic, and red pepper.

While gallivanting around, Pie and Scriber had been good er to grab a mountain chicken for me. This amazing, powerful speci. was the first live one I had seen! I admired its muscular body, noting that a very firm grip was required to hold it, but its brawn had secured its place on the menus of restaurants around the island.

This one was spared, but the heavy predation pressure by restaurants ensured that mountain chickens were getting difficult to find up in the forests in 1994 (approximately fifty adult frogs per day made it into a pot for consumption). During the volcanic crisis, there had been great concern for this species in light of the volcano-polluted environment and the skin lesions noted on many animals. The concern was so great that several trillion billion English pounds (I may be slightly exaggerating here) were spent to whisk away several breeding pairs of these big hoppers on an all-expenses-paid "honeymoon" to England in an attempt to "save" the species.

Many years later, I found this swashbuckling conservation effort ludicrous, given that these huge frogs had become so numerous along forest trails that I had to be somewhat careful as to where I stepped as I moved among my mist nets in the dark. It was obvious what had happened: the partial evacuation of humans from the island had significantly reduced hunting pressure, and the frog population rebounded quite dramatically of its own accord. As people move back to Montserrat, it will be interesting to see if this resurgence in the

Leptodactylus population stabilizes or if these frogs will once again find themselves back on dinner tables around Montserrat.*

Road-toad and Mud Rain

Karen Hadley had contacted me out of the blue in 1994 and volunteered to come down to Montserrat to catch bats with me. I arranged to pick her up at the airport after I got off work and had showed up a bit late. Never having met Karen, I wasn't sure who or what to expect as I pulled into the airport parking lot, but I soon noted a huge backpack moving toward me atop two elfin feet. As this apparition neared, I noticed that the little feet were attached to a huge grin, which was in turn attached to Karen.

Karen spent her days exploring Montserrat and soaking up all the island had to offer. We went out many times to resample the bats at several of my netting localities—we had a wonderful time. Sadly, it didn't end well. For lunch on her last day on island, Karen decided to try the local delicacy—mountain chicken—at a restaurant in Plymouth. To this day, we aren't entirely sure what she was served (road-toad—*Bufo marinus*—perhaps?), but she and her meal parted company shortly thereafter, leaving her in alternating shades of green not becoming a healthy mammal. She departed the next day, and I feared that I would never see Karen in the field again.

But in July 2001, my colleagues Karen Hadley, Gary Kwiecinski, and I settled into a tidy villa overlooking the Belham River valley in preparation for what we hoped would be a wonderful couple of weeks of fieldwork. The forest was very dry that year, but our capture rates were up. The general mood of my small band was relaxed and upbeat. Then it started to rain—a tropical wave had snuck up on us. Normally, this

---

* This query will never be answered. Chytrid fungus first appeared on Montserrat in 2009. It had decimated the mountain chicken population by 2010.

would mean that we would spend two or three days loafing around the house drinking beer, catching up on the local hit tunes on the radio, playing cards, repairing damaged mist nets, and cleaning up our field notes and species accounts. But on the second day, the volcano apparently had had enough of this stormy nonsense. Its eastern flank had absorbed so much water from the storm that several landslides exposed the hot core of the mountain. The result was a spectacular eruption of steam, rocks, and ash on the afternoon of July 29.

The cool hard rains began to change, imperceptibly at first, into warm splattering mud droplets. Though others might have exercised a bit more caution, we piled into the Jeep around 3:00 p.m. and drove down into the valley to see what was going on. We found that roiling floodwaters had cut the road and were flushing medium-sized trees and debris across the flat valley floor out to the ocean where fifty yards or so of new beachfront property had been just been created out beyond the old surf line. Despite our macabre fascination with the flood and the black rain, we eventually exercised a modicum of common sense and returned to our house. We spent the next hour boarding up the place to the best of our ability, and I made some calls to colleagues at the north end of the island. They hadn't heard anything on the radio about any renewed volcanic activity. (The government is funny that way—it provides a "warning" to people that a pyroclastic flow had occurred the day *after* the event actually happened.) But no worries, right? The power was still on (we had several cases of beer in the refrigerator) and the phone and TV still worked. Then the wind changed direction. Horizontal mud-rain and pebbles filled the air and the whole world came down on top of us, or so it seemed. The power went out, the sky turned black, and rocks one to three inches in diameter began to rain down on the roof of our house with a deafening racket. (I use one of these rocks as a paperweight on my desk to this day.)

We had little choice but to sit and wait. The rocks stopped falling around 5:00 p.m., and I went outside with my camcorder to film the single most horrifying and spectacular sight I'd seen in my life. The volcanic eruption towered above my head to well over 20,000 feet in a huge snake-like plume that arced up from the volcano and headed out to sea. Flanked by blinding-blue clear sky to its north, the beast writhed and heaved in agony back and forth across the evening sky far above our house, with each gyration pelting us with gravel and plastering us with another half inch or so of wet concrete-like mud.

I can't speak for Karen or Gary, but I felt very small indeed.

That night was difficult. After a brief respite, the tropical wave returned to buffet us with high winds and dropped great sheets of rain on us, intermixed with clouds of volcanic ash. We spent the night under our sheets thinking our own thoughts and trying to filter out as much ash as we could with wet towels and bandanas—Karen being the most stalwart of the group. I went outside several times to dig diversion channels through the ash and silt to keep water from flooding through the kitchen and was fascinated to see how quickly the extremely fine-grained muck could be washed away by rivulets of rain.

By morning, the four to five inches of loosely compacted ash had been reduced to a thick sludge of half an inch to one and a half inches on all horizontal and vertical surfaces, as if splatter-painted by a huge spray gun. We emerged as pale ghosts into a grim black and gray world devoid of all color and sound. Nothing moved. We needed the low range of our 4WD to get out of our driveway. As we drove north, we were slowly revived by the beautiful colors and luxuriant greens of a tropical forest glistening with a newness that can only be seen after a good rain. It was as if we had been spit out onto another planet.

We were starving. We went to a restaurant where our waitress dismissively commented, "Oh! Did it ash last night?" After what we had gone through, I was incredulous. Until I realized with great humility that she and the entire island had been beset by even more grotesque volcanic horrors throughout the previous four years. What we had experienced was insignificant by her standards. But according to the Montserrat Volcanic Observatory, it had been a major eruption, one of the largest to date, and thankfully it had impacted only two abandoned villages and three bat biologists.

With respect to our bat project that season, we had already surveyed our northern mist netting sites, having saved my favorite southern sites (now buried in ash) for the last week of the trip. We could have tried to continue our work by trudging through the mud or scuffing through suffocating ash, but the writing was on the wall. We bugged out. We did our best to hitch rides on those great silver freedom-birds—Karen found direct flights, while me, myself, and I spent seventy-two hours eating and sleeping in airport bars (Antigua, Puerto Rico, Miami, St. Louis, Minneapolis), filling my notebook with copious half sentences and half-completed thoughts in an attempt to record all that we had seen and experienced.

While writing this chapter, I discovered that a fine dusting of volcanic ash remains on my notebook pages.

Diversity and Disasters and Dentifrice, Oh, My!

The dark green forests that once covered the flanks of Montserrat's volcano provided the island not only with its nickname—the Emerald Isle—but with habitat that supported a remarkable level of bat diversity for an island of this size in this corner of the world (five fruit bats; one nectarivore; three insectivores; and the fishing bat, *Noctilio*).

Possibly because of the absence of white sand beaches and a deep harbor, Montserrat has dodged the saprophytic tourism industry, but it has received a great deal of unappreciated attention from Mother Nature. The island has been battered by more than thirty hurricanes in the past 360 years. Since 1997, the volcano has busied itself burying both villages and river drainages alike under meters of sterile volcanic ash. The fig tree from hell was destroyed by a pyroclastic flow in 1997. It has been heartbreaking for me to revisit old netting sites that I had known as lush valleys, only to find nightmarish visions of what looked like the surface of the moon.

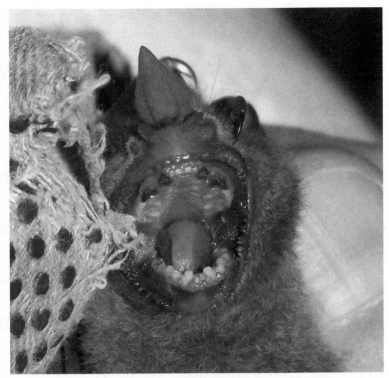

8.8. Montserrat fruit-eating bat's blackened and damaged teeth from eating ash-covered vegetation (see figure 8.4 for comparison)

Fruit bat populations on Montserrat have fluctuated dramatically over the years. In this disturbance-driven ecology, I have had to rethink what the phrase "rare species" means. However, it was my training as an anatomist that brought me to Montserrat in the first place—population biology is far less interesting to me than recording the physiological wear and tear incurred by the bats that operate in damaged habitat. As one might predict, major eruptive cycles are mirrored by the appearance of several sublethal pathologies in these animals, including emphysema, silicoproteinosis, nephritis, increased parasite loads, idiopathic baldness . . . and severe tooth wear.

For those readers who are obsessed with their own teeth, volcanic ash is a fine abrasive often used as a dentifrice. However, after a pyroclastic eruption, this grit covers everything, and it is next to impossible for a fruit bat to avoid ingesting this nasty stuff during feeding or grooming. The abrasive material eventually destroys the teeth. As such, it is quite easy to distinguish gummy old bats from young bats that have never encountered ash simply by offering them an exposed thumb—but here my story has come full circle and I digress.

Through a Beer Glass Darkly

For twenty years, I have scribbled away at my field notes immediately after I have taken down my nets—typically in a small neighborhood bar while my memories are still fresh . . . initially. For this essay, rereading these notes was a poignant reminder of how fortunate I have been to observe bat populations and how they respond to natural disasters. Despite the minor inconveniences of being blown out to sea by hurricanes or incinerated by pyroclastic flows, the bats of Montserrat have soldiered on . . . encouraging me to continue stumbling around in the forests of Montserrat with my nets and headlamp. There is so much to learn about this troubled ecosystem, yet I quite expect to remain in the dark on this subject, quite literally—after all, that's where the bats are.

Acknowledgments. Heartfelt acknowledgment must be made to Dave Armstrong for instilling in me his love of mammals and for giving me a tremendous break that altered my life forever, and to Hugh Genoways for doing me the great honor of being my friend and colleague.

Contributors

Dr. Rick A. Adams is professor of biology at the University of Northern Colorado. He has studied bats and other mammals for more than twenty years in the Rocky Mountain West and the Caribbean islands. He is the author of *Bats of the Rocky Mountain West: Natural History, Ecology, and Conservation* and coeditor of *Ontogeny, Functional Ecology, and Evolution of Bats*. His work has been published frequently in scientific journals and in books as well as *Natural History* magazine. Rick's research concentrates on the population and community ecology of bats in western North America, the Caribbean islands, and South Africa. He is founder and president of the Colorado Bat Society, which works to conserve bats and their habitats in the West. He currently lives in Boulder.

Christina Allen received her master of science degree in tropical conservation and development from the University of Florida in 1996. Her research focusing on forest fragmentation in the Peruvian Amazon was featured in National Geographic's book *Talking with Adventurers*. She is author of *Hippos in the Night: Autobiographic Adventures in Africa*. Her experiences include diving in the Galapagos as a member of the GalapagosQuest team, her first job out of graduate school. As a team member, she traveled cross-country, mainly by bicycle, through many countries, teaching schoolchildren about science and nature via the Internet while on "live expeditions." Christina grew up in Anchorage and currently resides in Boulder.

Dr. Frank J. Bonaccorso received his Ph.D. in zoology from the University of Florida in 1975. He has conducted research on mammals, birds, plants, and insects; his study locations include South Africa, Papua New Guinea, Venezuela, Costa Rica, Panama, Ecuador, and Belize. After serving many years as chief curator of natural history at the Papua New Guinea National Museum, he is currently a wildlife biologist with the U.S. Geological Survey in Hawaii, conducting research on endangered bats. He is author of *Bats of Papua New Guinea* as well as many publications in scientific journals. Frank currently resides two miles from the caldera of Kilauea Volcano in Volcano Village on Big Island of Hawaii.

Dr. Lee Dyer is an ecologist who has worked with a variety of organisms in the tropics for the past fifteen years and in temperate areas for the past nineteen years. His research examines interactions among plants, herbivores, and their natural enemies and includes work in Costa Rica, Colorado, and California. Lee was a professor for five years at Mesa State College in Colorado, where he established the Western Colorado Center for Tropical Research and received the distinguished faculty scholar award. In 2001, Lee became a faculty member in the ecology and evolutionary biology department at Tulane University. His academic specialties are tropical biology, statistical modeling, community ecology, caterpillar natural history, and natural products chemistry. He has published many scientific articles on his research. He is currently in the biology department at the University of Nevada–Reno.

Dr. James C. Halfpenny owns A Naturalist's World, a company dedicated to providing educational programs, books, slide shows, and videos about ecologically important subjects. Since 1961, Jim has taught outdoor education and environmental programs for state, federal, and private organizations. Jim is a fellow of the Explorer's Club and has led expeditions in Antarctica, China, Greenland, Kenya, Tanzania, and the United States. He is the author of *Yellowstone Wolves in the Wild*, *A Field Guide to Mammal Tracking in North America*, *Winter: An Ecological Handbook*, the series Scats and Tracks, and *Yellowstone Bears in the Wild*. He has published numerous scientific and popular articles. He resides in Gardiner, Montana, doorway to the Greater Yellowstone Ecosystem.

Stephen R. Jones is author of *The Last Prairie, a Sandhills Journal* and coauthor, with Ruth Carol Cushman, of the *Peterson Field Guide to the North American Prairie*, *The Shortgrass Prairie*, and *Colorado Nature Almanac*. He taught in the Boulder Valley Public Schools for thirty-three years and now consults and teaches field ecology classes for the Boulder County Nature Association. His research in the Nebraska Sandhills focuses on the effects of environmental change on breeding bird populations. The place he calls "Pine Lake" in his essay is given a different name on modern maps. He currently lives in Boulder.

Dr. Ann Kohlhaas is a professor in the Department of Biology at California State University, Stanislaus. Her research focuses on the ecology, behavior, and conservation of vertebrates, especially primates but including other mammals and birds. She has worked extensively in the Malay Archipelago on the island of Sulawesi in Indonesia, where she studies the ecology and behavior of macaque monkeys. Ann currently resides in Turlock, California.

Dr. Scott C. Pedersen is associate professor of biology at South Dakota State University (SDSU), where he teaches courses in gross anatomy, evolution, and embryology. He is also curator of mammals in the Natural History Collections at SDSU and is a research associate of the Museum of Texas Tech University and the University of Nebraska State Museum. Scott is coeditor of *Ontogeny, Functional Ecology and Evolution of Bats* and has many publications in journals and books, including *Natural History* magazine. He has spent the past fifteen years chasing bats throughout the Caribbean islands with his graduate students and native wildlife officials. He has documented that the Soufrière Hills volcano on Montserrat has made unique and significant impacts on the island's bat populations. He currently lives in Brookings, South Dakota.